D0123249

Other Monographs in this Series

No. 1. Irving Kaplansky: *Algebraic and analytic aspects of operator algebras*

2. Gilbert Baumslag: *Lecture notes on nilpotent groups*

3. Lawrence Markus: *Lectures in differentiable dynamics*

4. H. S. M. Coxeter: *Twisted honeycombs*

5. George W. Whitehead: *Recent advances in homotopy theory*

6. Walter Rudin: *Lectures on the edge-of-the-wedge theorem*

7. Yozo Matsushima: *Holomorphic vector fields on compact Kähler manifolds*

8. Peter Hilton: *Lectures in homological algebra*

9. I. N. Herstein: *Notes from a ring theory conference*

10. Branko Grünbaum: *Arrangements and spreads*

11. Irving Glicksberg: *Recent results on function algebras*

12. Barbara L. Osofsky: *Homological dimensions of modules*

13. Michael Rabin: *Automata on infinite objects and Church's problem*

14. Sigurdur Helgason: *Analysis on Lie groups and homogeneous spaces*

15. R. G. Douglas: *Banach algebra techniques in the theory of Toeplitz operators*

16. Joseph L. Taylor: *Measure algebras*

17. Louis Nirenberg: *Lectures on linear partial differential equations*

18. Avner Friedman: *Differential games*

19. Béla Sz.-Nagy: *Unitary dilations of Hilbert space operators and related topics*

20. Hyman Bass: *Introduction to some methods of algebraic K-theory*

Conference Board of the Mathematical Sciences
REGIONAL CONFERENCE SERIES IN MATHEMATICS

supported by the

National Science Foundation

Number 21

HOLOMORPHIC FUNCTIONS OF FINITE ORDER IN SEVERAL COMPLEX VARIABLES

by

WILHELM STOLL

Published for the

Conference Board of the Mathematical Sciences

by the

American Mathematical Society

Providence, Rhode Island

Expository Lectures
from the CBMS Regional Conference
held at the University of Wisconsin, Watewater
June 18–22, 1973

AMS (MOS) Subject Classification (1970): 32A15

Library of Congress Cataloging in Publication Data

CIP

Stoll, Wilhelm.
Holomorphic functions of finite order in several complex variables.
(Regional conference series in mathematics, no. 21)
"Expository lectures from the CBMS regional conference held at the University of Wisconsin, Whitewater, June 18–22, 1973."
1. Analytic functions. 2. Functions, Entire.
I. Conference Board of the Mathematical Sciences.
II. Title. III. Series.
QA1.R33 no. 21 [QA331] 510′.8s [515′.92′5] 74-8213
ISBN 0-8218-1671-3

Copyright © 1974 by the American Mathematical Society

Printed in the United States of America

All rights reserved except those granted to the United States Government.
This book may not be reproduced in any form without permission of the publishers.

1413853

MATH.-SCI.

QA
331
.S77

Math Sci
Sep

CONTENTS

Dedicated to the founders of this field in several variables.

Stefan Bergman Hellmuth Kneser Pierre Lelong

Preface

From June 18 to 22, 1973 a National Science Foundation Regional Conference on holomorphic functions of finite order in several variables was held at the University of Wisconsin-Whitewater, Wisconsin. In one variable, the theory of functions of finite order is well developed and a number of expositions exist. In several variables, the theory has not reached this stage of maturity but the basic principles and ideas are emerging.

The allotment of ten hours of lecture during the conference week did not permit to survey in depth all the principal features of the theory. Therefore one fundamental aspect was selected as the central topic of this survey. As such I selected the construction of holomorphic functions with growth estimates to given zero sets. This topic is relatively easily accessible and has many similarities to the one variable theory. A coherent account is lacking. Important new results have been obtained and the recall of older results is of value. Other aspects are better covered. For instance, during the spring of 1973 a special semester in value distribution in several variables was held at Tulane University. The proceedings will provide a wide panorama of the investigations into the properties of a holomorphic map. In view of these reasons this monograph will concentrate on the construction of holomorphic functions with growth estimates to given zero sets. As an exception to this rule, the two main theorems of value distribution will be stated here for a meromorphic map of a hermitian vector space into a complex projective space, because of the fundamental importance of these theorems. After three preparatory sections, this will be done in §4.

In §5, the Jensen-Poisson formula for balls will be proved and applied to the construction of a canonical function on a relatively compact ball in the domain of the given divisor. In §6, the canonical function for a divisor in a hermitian vector space will be constructed using weights. The connection to the canonical function of Lelong [28] is established; however, I will not pursue the theory of Lelong and his school further. For a survey see Lelong [29], [30]. The results on the canonical function are applied to construct theta functions to periodic divisors. Rubel and Taylor [36] had introduced a Fourier series method to study and construct functions of finite λ-type in one variable. Kujala [23] extended this investigation to several variables. A short survey of his results is given in §8. Time limitations did exclude this topic from being presented at the conference. In §9, the canonical function of a divisor of finite order on the unit ball is introduced and a factorization theorem is given. §9 reviews results contained in the dissertation of Mueller [31]. Again, consider the case of a nonnegative divisor ν on a hermitian vector space. Ronkin [35] gives a third integral representation of the canonical function of ν. In §10, Ronkin's results will be completely restructured and considerably advanced. Ronkin uses a power series method similar to Kneser's. Here a different method will be used. A Jensen-Poisson formula for polydiscs will be established and analogues to the Weierstrass product and the Blaschke product will be given. An integral representation of a good inner function on a polydisc will be obtained. Since a good part of §10 is new, complete proofs will be given here. In §11, a survey of fundamental results of Skoda [44] will be given. Skoda uses ingenious new ideas, extensive calculations and deep results of Hörmander.

Hopefully, this survey will make the topic better known and easier accessible and will, in consequence, stimulate new research into the many open questions remaining.

I would like to express my gratitude to Professor Rudolph Najar for his kind invitation and his devoted efforts in organizing this conference.

WILHELM STOLL
UNIVERSITY OF NOTRE DAME

1. Review of Some Basic Results in One Variable

The complex projective space $\mathbf{P}_1$ of dimension one can be considered as the Riemann sphere, embedded into $\mathbf{R}^3$ as a sphere of diameter 1. For any two points a and w in $\mathbf{P}_1$, denote by $\|w; a\|$ the length of the line segment determined in $\mathbf{R}^3$ by the edges a and w. Obviously $0 \leqslant \|w; a\| \leqslant 1$. Let $\ddot{\omega}$ be the rotation invariant volume element of $\mathbf{P}_1$ in $\mathbf{R}^3$ normalized such that $\int_{\mathbf{P}_1} \ddot{\omega} = 1$. The Riemann sphere $\mathbf{P}_1$ is a compact, complex manifold. As on any complex manifold, the exterior derivative on differential forms splits into $d = \partial + \bar{\partial}$ and twists to $d^c = i(\bar{\partial} - \partial)$. Then

$$\ddot{\omega}(w) = \frac{1}{2\pi} dd^c \log \frac{1}{\|w; a\|}$$

on $\mathbf{P}_1 - \{a\}$.

Let f be a nonconstant meromorphic function on $\mathbf{C}$. Then f can be considered as a holomorphic map

$$f : \mathbf{C} \longrightarrow \mathbf{P}_1.$$

The map f spreads the disk $\mathbf{C}(r) = \{z \in \mathbf{C} |\ |z| > r\}$ over $\mathbf{P}_1$. The volume of this spread is given by

$$A_f(r) = \int_{\mathbf{C}(r)} f^*(\ddot{\omega}) \geqslant 0.$$

The characteristic of f is defined for $0 < s < r < +\infty$ by

$$T_f(r, s) = \int_s^r A_f(t) \frac{dt}{t}.$$

Let σ_r be the invariant measure of the circle $\mathbf{C}\langle r \rangle = \{z \mid |z| = r\}$ such that $\int_{\mathbf{C}\langle r \rangle} \sigma_r = 1$. For each $a \in \mathbf{P}_1$ and $r > 0$, the compensation function of f is defined by

$$m_f^a(r) = \int_{\mathbf{C}\langle r \rangle} \log \frac{1}{\|f; a\|} \sigma_r \geqslant 0.$$

For each $a \in \mathbf{P}_1$ and $z \in \mathbf{C}$, the multiplicity $\mu_f^a(z) \geqslant 0$ is defined with which f assumes the value a at z. The a-divisor of f is defined as the function

$$\mu_f^a : \mathbf{C} \longrightarrow \mathbf{Z}.$$

Define $\mu_f = \mu_f^0 - \mu_f^\infty$ as the divisor of f. A divisor ν on $\mathbf{C}$ is any function $\nu : \mathbf{C} \to \mathbf{Z}$ such that $\mu_f = \nu$ for some meromorphic function on $\mathbf{C}$. If ν is a divisor on $\mathbf{C}$, define the counting function for all $r \geqslant 0$ by

$$n_\nu(r) = \sum_{|z| \leqslant r} \nu(z).$$

The valence function of ν is defined by

$$N_\nu(r, s) = \int_s^r n_\nu(t) \frac{dt}{t}.$$

If $\nu = \mu_f^a$, write

$$N_\nu = N_f^a, \qquad n_\nu = n_f^a.$$

If $\nu = \mu_f$, write

$$N_\nu = N_f, \qquad n_\nu = n_f.$$

The First Main Theorem asserts

$$T_f(r, s) = N_f^a(r, s) + m_f^a(r) - m_f^a(s)$$

for all r and s with $0 < s < r$ and all $a \in \mathbf{P}_1$. Observe $T_f(r, s) \to \infty$ for $r \to \infty$. Hence

$$0 \leqslant \delta_f(a) = \liminf_{r\to\infty} \frac{m_f^a(r)}{T_f(r, s)} = 1 - \limsup_{r\to\infty} \frac{N_f^a(r, s)}{T_f(r, s)} \leqslant 1$$

is independent of $s \geqslant 0$ and is called the defect of f at a.

Let $M \neq \emptyset$ be a finite subset of $\mathbf{P}_1$. The Second Main Theorem asserts

$$\sum_{a \in M} \delta_f(a) \leqslant 2.$$

If $f^{-1}(a) = \emptyset$, then $\delta_f(a) = 1$. If f is not rational, and if $f^{-1}(a)$ is finite, then $\delta_f(a) = 1$. Hence there can be at most two such values a in these cases. Particularly, the Picard's theorem is implied.

For every $z \in \mathbf{C}$ and $0 \leqslant q \in \mathbf{Z}$ define $\exp(z) = e^z$ and

$$E(z, q) = (1 - z) \exp\left(\sum_{\mu=1}^{q} \frac{z^\mu}{\mu} \right).$$

Let $0 \leqslant \nu : \mathbf{C} \to \mathbf{Z}$ be a nonnegative divisor. Define $\mathbf{R}_+ = \{x \in \mathbf{R} \mid x \geqslant 0\}$. An integral valued, increasing function $q \geqslant 0$ on $\mathbf{R}_+$ is called a weight for ν if and only if

$$\sum_{0 \neq z} \nu(z)\, (r/|z|)^{q(|z|)+1} < \infty$$

converges for each $r > 0$. A weight function always exists. If q is a weight function, then the product

$$h(z) = z^{\nu(0)} \prod_{y \neq 0} E(z/y, q(|y|))^{\nu(y)}$$

converges uniformly on compact subsets of $\mathbf{C}$ and defines an entire function with $\mu_h^0 = \mu_h = \nu$. If there exists a constant weight function q for ν, then h is a function of finite order and precise growth estimates can be made.

For instance, let $w_1 \in \mathbf{C}$ and $w_2 \in \mathbf{C}$ be linearly independent over $\mathbf{R}$. Define $\nu(z) = 1$ if $z = n_1 w_1 + n_2 w_2$ with integers n_1 and n_2 and define $\nu(z) = 0$ otherwise. Then $\nu \geqslant 0$ is a divisor and $q = 2$ is a constant weight function. The associated canonical function h is the sigma function of Weierstrass.

Similar results hold for the unit disc $\mathbf{C}(1)$. For $z \in \mathbf{C}(1)$ and $0 \neq y \in \mathbf{C}(1)$ and $0 \leqslant q \in \mathbf{Z}$ define

$$b(z, y) = \frac{1 - z/y}{1 - z\bar{y}}, \qquad b(z, y; q) = E(1 - b(z, y); q).$$

Observe $b(z, y; 0) = b(z, y)$.

Let $\nu \geqslant 0$ be a divisor on the unit disc. Let $q \geqslant 0$ be an integer. Assume that

$$\sum_{z \in \mathbf{C}(1)} \nu(z)\,(1 - |z|)^{q+1} < \infty \tag{1.1}$$

converges. Then

$$B(z) = z^{\nu(0)} \prod_{0 \neq y \in \mathbf{C}(1)} b(z, y; q)^{\nu(y)}$$

converges uniformly on each compact subset of $\mathbf{C}(1)$ and defines a holomorphic function on the unit disc with $\mu_B^0 = \mu_B = \nu$. The growth of B can be estimated. This result of Mueller [31] is a variation of an older result of Tsuij [64].

For $q = 0$, condition (1.1) is equivalent to the Blaschke condition $N_\nu(1, s) < \infty$ for some s with $0 < s < 1$. If $\nu(0) = 0$, then $s = 0$ is permitted and $B \exp(-N_\nu(1))$ is the Blaschke product.

Fundamental to the derivation of these results is the Jensen-Poisson formula:

Let $0 < R \leqslant +\infty$. Let $f \not\equiv 0$ be a meromorphic function on the disc $\mathbf{C}(R) = \{z \mid |z| < R\}$ such that $f(0) \neq 0$ and $f(0) \neq \infty$. Take any r with $0 < r < R$. Let G by any open, connected, simply connected neighborhood of 0 which is contained in $C(r) - \operatorname{supp} \nu$. Take $z \in \mathbf{C}(r)$. Then

$$\log \frac{f(z)}{f(0)} = \frac{1}{2\pi} \int_0^{2\pi} \log |f(re^{i\phi})| \frac{2z}{re^{i\phi} - z}\, d\phi$$

$$+ \sum_{|u| < r} \mu_f(u) \log \left(1 - \frac{z}{u}\right) - \sum_{|u| < r} \mu_f \log \left(1 - \frac{z\bar{u}}{r^2}\right),$$

$$\log |f(0)| = \frac{1}{2\pi} \int_0^{2\pi} \log |f(re^{i\phi})|\, d\phi - N_f(r).$$

Here, as everywhere in this monograph, the following convention shall be observed: Let G be an open, connected, simply connected neighborhood of the origin 0 of a complex vector space. Let f be a holomorphic function on G with $f(0) > 0$. Then $\log f$ is understood to be that holomorphic function $F = \log f$ uniquely defined by $F(0) \in \mathbf{R}$ and $f = e^F$.

At this point, the review of the one variable case shall be terminated. Almost all of the mentioned results in one variable can be carried over to several variables in one or several fashions. Some of these extensions and the associated techniques shall be reviewed in this monograph.

2. Hermitian Vector Spaces

The geometric and differential geometric structures involved in several variables is considerably richer than in the one variable case. Therefore, two sections of preparations are needed to introduce the relevant notations and concepts.

If S is a set, let S^n be the n-fold cartesian product and let $\#S$ be the number of elements in S. If S is partially ordered, define

$$S[a, b] = \{x \in S \mid a \leqslant x \leqslant b\},$$

$$S(a, b] = \{x \in S \mid a < x \leqslant b\},$$

$$S[a, b) = \{x \in S \mid a \leqslant x < b\},$$

$$S(a, b) = \{x \in S \mid a < x < b\}.$$

Here, a and b do not have to belong to S. Only the notation must be meaningful. For example,

$$\mathbf{Z}_+ = \mathbf{Z}[0, +\infty) = \{x \in \mathbf{Z} \mid 0 \leqslant x\},$$

$$\mathbf{R}_+ = \mathbf{R}[0, +\infty) = \{x \in \mathbf{R} \mid 0 \leqslant x\},$$

$$\mathbf{R}^+ = \mathbf{R}(0, \infty) = \{x \in \mathbf{R} \mid 0 < x\}.$$

A hermitian vector space W is a complex vector space together with a positive definite hermitian form on W. The hermitian product of $\mathfrak{a} \in W$ and $\mathfrak{b} \in W$ is denoted by $(\mathfrak{a} \mid \mathfrak{b})$. The norm of $\mathfrak{a}$ is $|\mathfrak{a}| = \sqrt{(\mathfrak{a} \mid \mathfrak{a})}$.

Throughout this monograph, W shall denote a hermitian vector space of finite dimension m with $m > 0$. For each $p \in \mathbf{N}[1, m]$, the exterior product $\bigwedge_p W$ is a hermitian vector space such that

$$(\mathfrak{a}_1 \wedge \cdots \wedge \mathfrak{a}_p \mid \mathfrak{b}_1 \wedge \cdots \wedge \mathfrak{b}_p) = \det (\mathfrak{a}_\mu \mid \mathfrak{b}_\nu).$$

The dual vector space W^* of W is the vector space of all linear functions $\alpha : W \to \mathbf{C}$. For each $\alpha \in W^*$, one and only one $\mu(\alpha) \in W$ exists such that $\alpha(\mathfrak{z}) = (\mathfrak{z} \mid \mu(\alpha))$ for all $\mathfrak{z} \in W$. The map $\mu : W^* \to W$ is an antilinear isomorphism. One and only one hermitian product exists on W^* such that

$$(\alpha \mid \beta) = (\mu(\beta) \mid \mu(\alpha)).$$

Two vectors $\mathfrak{a} \neq 0$ and $\mathfrak{b} \neq 0$ in W are said to be equivalent if and only if $\mathfrak{a} = \lambda \mathfrak{b}$ for some $\lambda \in \mathbf{C}$. This defines an equivalence relation on $W - \{0\}$. Let $\mathbf{P}(\mathfrak{z})$ be the equivalence class of $\mathfrak{z} \in W - \{0\}$. If $A \subseteq W$, define

$$\mathbf{P}(A) = \{P(\mathfrak{z}) \mid 0 \neq \mathfrak{z} \in A\}.$$

Then $\mathbf{P}(W)$ is a connected, compact complex manifold of dimension $m-1$, called the complex projective space of W such that $\mathbf{P}: W - \{0\} \to \mathbf{P}(W)$ is holomorphic.

Take $p \in \mathbf{Z}[0, m)$. The Grassmann cone of p is defined by

$$\widetilde{G}_p(W) = \{\mathfrak{a}_0 \wedge \cdots \wedge \mathfrak{a}_p \mid \mathfrak{a}_\mu \in W\}.$$

The Grassmann manifold for p is defined by

$$G_p(W) = \mathbf{P}(\widetilde{G}_p(W)) \subseteq \mathbf{P}(\textstyle\bigwedge_{p+1} W)$$

which is a connected, compact, smooth, complex submanifold of $\mathbf{P}(\bigwedge_{p+1} W)$ with

$$d(m, p) = \dim G_p(W) = (p+1)(m-p-1)$$

and with degree

$$\Delta(m, p) = d(m, p)! \prod_{\mu=1}^{m} \frac{\mu!}{(m-1-\mu)!}.$$

If $0 \neq \mathfrak{a} \in \widetilde{G}_p(W)$, then $E(\mathfrak{a}) = \{\mathfrak{z} \in W \mid \mathfrak{z} \wedge \mathfrak{a} = 0\}$ is a linear subspace of dimension $p+1$ of W. If $\mathfrak{a} = \mathfrak{a}_0 \wedge \cdots \wedge \mathfrak{a}_p$, then $\mathfrak{a}_0, \cdots, \mathfrak{a}_p$ is a base of $E(\mathfrak{a})$ over $\mathbf{C}$. Then the complex projective space

$$\ddot{E}(\mathfrak{a}) = \mathbf{P}(E(\mathfrak{a}))$$

of $E(\mathfrak{a})$ is smoothly embedded into $\mathbf{P}(W)$ and is called a p-dimensional projective plane in $\mathbf{P}(W)$. If $0 \neq \lambda \in \mathbf{C}$, then $E(\lambda\mathfrak{a}) = E(\mathfrak{a})$ and $\ddot{E}(\lambda\mathfrak{a}) = \ddot{E}(\mathfrak{a})$. Thus, if $a \in G_p(W)$, take any $\mathfrak{a} \in P^{-1}(a)$; then

$$E(a) = E(\mathfrak{a}), \qquad \ddot{E}(a) = \ddot{E}(\mathfrak{a}) = \mathbf{P}(\ddot{E}(a))$$

are well defined.

Take $0 \neq \alpha \in W^*$. Then $E[\alpha] = \ker \alpha$ is a linear subspace of dimension $m-1$ in W. Then

$$\ddot{E}[\alpha] = \mathbf{P}(E[\alpha]) \subseteq \mathbf{P}(W)$$

is a projective complex plane of dimension $m-2$ in $\mathbf{P}(W)$ also called a hyperplane. If $0 \neq \lambda \in \mathbf{C}$, then $E(\lambda\alpha) = E(\alpha)$ and $\ddot{E}(\lambda\alpha) = \ddot{E}(\alpha)$. If $a \in \mathbf{P}(W^*)$, take any $\alpha \in \mathbf{P}^{-1}(a)$; then

$$E[a] = E[\alpha], \qquad \ddot{E}[a] = \ddot{E}[\alpha] = \mathbf{P}(E[a])$$

are well defined.

Given $a \in \mathbf{P}(W^*)$ and $w \in \mathbf{P}(W)$, take any $\alpha \in \mathbf{P}^{-1}(a)$ and $\mathfrak{w} \in \mathbf{P}^{-1}(w)$. Then

$$0 \leqslant \|a; w\| = \|w; a\| = |\alpha(\mathfrak{w})|(|\alpha|\ |\mathfrak{w}|) \leqslant 1$$

is well defined with $\ddot{E}[a] = \{w \in \mathbf{P}(W) \mid \|a; w\| = 0\}$.

The projective space $\mathbf{P}(W)$ is a Kaehler manifold with a fundamental form $\ddot{\omega} > 0$ of bidegree $(1, 1)$ of the Kaehler metric uniquely defined such that

$$\ddot{\omega}(w) = \frac{1}{2\pi}\, dd^c \log \frac{1}{\| w; a \|}$$

on $\mathbf{P}(W) - \ddot{E}[a]$ for each $a \in \mathbf{P}(W^*)$. The Kaehler metric is normalized such that $\int_{\mathbf{P}(W)} \ddot{\omega}^{m-1} = 1$. This Kaehler metric is called the Fubini-study metric. Observe that $\mathbf{P}(W^*)$ is a Kaehler manifold whose fundamental form is denoted by $\ddot{\omega}$ again. Also $\mathbf{P}(\bigwedge_{p+1} W^*)$ is a Kaehler manifold whose fundamental form is denoted by $\ddot{\omega}_p$. Then

$$\Delta(m, p) = \int_{G_p(W)} \ddot{\omega}_p^{d(m,p)}.$$

Let us return to the vector space W. If $A \subseteq W$ and $r \geqslant 0$, define

$$A[r] = \{ \mathfrak{z} \in A \mid |\mathfrak{z}| \leqslant r\},$$

$$A(r) = \{ \mathfrak{z} \in A \mid |\mathfrak{z}| < r\},$$

$$A\langle r\rangle = \{ \mathfrak{z} \in A \mid |\mathfrak{z}| = r\},$$

$$\tau_0 : W \longrightarrow \mathbf{R} \quad \text{by} \quad \tau_0(\mathfrak{z}) = |\mathfrak{z}|,$$

$$\tau = \tau_0^2 : W \longrightarrow \mathbf{R} \quad \text{by} \quad \tau(\mathfrak{z}) = |\mathfrak{z}|^2.$$

Define the differential forms

$$\upsilon = \frac{1}{4\pi}\, dd^c\tau > 0 \qquad \text{on } W,$$

$$\rho = \frac{1}{4\pi}\, d^c\tau \wedge \upsilon^{m-1} \qquad \text{on } W,$$

$$\omega = \frac{1}{4\pi}\, dd^c \log \tau \geqslant 0 \qquad \text{on } W - \{0\},$$

$$\sigma = \frac{1}{4\pi}\, d^c \log \tau \wedge \omega^{m-1} \qquad \text{on } W - \{0\}.$$

Obviously,

$$d\rho = \upsilon^m, \qquad d\sigma = \omega^m, \qquad \sigma = \rho\tau^{-m}.$$

Here, υ^p is the euclidean volume element on each p-dimensional complex submanifold of W normalized such that $\int_{W(r)} \upsilon^m = r^{2m}$.

Here σ pulls back to the euclidean volume element on each sphere $W\langle r\rangle$ normalized such that $\int_{W\langle r\rangle} \sigma = 1$.

Here, ω is the projective differential such that $\omega = \mathbf{P}(\ddot{\omega})$.

Because $\mathbf{P}(W)$ has dimension $m - 1$, this implies $\ddot{\omega}^m = 0$. Hence

$$\omega^m = 0, \qquad d\sigma = 0.$$

The following identities are easily established:

$$\omega = \frac{\upsilon}{\tau} - \frac{1}{4\pi}\, \frac{d\tau \wedge d^c\tau}{\tau^2}, \qquad \omega^p = \frac{\upsilon^p}{\tau^p} - \frac{p}{4\pi}\, d\tau \wedge d^c\tau \wedge \frac{\upsilon^{p-1}}{\tau^{p+1}}.$$

Hence, $\omega^m = 0$ implies $4\pi\tau\upsilon^m = m d\tau \wedge d^c\tau \wedge \upsilon^{m-1}$. Let Δ be the Laplace operator on W. Let ψ be a function of class C^2 on some open subset of W. Then

$$m dd^c \psi \wedge \upsilon^{m-1} = \pi \Delta \psi \upsilon^m .$$

Therefore

$$\begin{aligned}\pi(\Delta\tau^{1-m})\upsilon^m &= m dd^c\tau^{1-m} \wedge \upsilon^{m-1}\\ &= m(1-m)d(\tau^{-m}d^c\tau \wedge \upsilon^{m-1})\\ &= m(1-m)\,d\sigma = 0\end{aligned}$$

which implies the well-known identity $\Delta\tau^{1-m} = 0$ on $W - \{0\}$. The identity $\omega^m = 0$ is not only important for elliptic differential equations, but is also the decisive fact which makes value distribution possible.

LEMMA 2.1. *Let $F : \mathbf{P}(W) \to \mathbf{C}$ be a measurable function. Take $r > 0$. Assume that at least one of the three integrals exists. Then all three integrals exist and*

$$\frac{1}{m!}\int_W (F \circ \mathbf{P})e^{-\tau}\upsilon^m = \int_{W\langle r\rangle} (F \circ \mathbf{P})\,\sigma = \int_{\mathbf{P}(W)} F\ddot{\omega}^{m-1}.$$

(See [46], [53].)

LEMMA 2.2. *Take $r > 0$. Let $F : W\langle r\rangle \to \mathbf{C}$ be a function. Assume $F\sigma$ is integrable over $W\langle r\rangle$. Then*

$$\int_{W\langle r\rangle} F\sigma = \int_{W\langle 1\rangle}\left[\frac{1}{2\pi}\int_0^{2\pi} F(re^{i\phi}\mathfrak{a})\,d\phi\right]\sigma(\mathfrak{a}).$$

(See [60], [53].)

If $W = \mathbf{C}^m$, then the hermitian metric shall always be given by

$$(\mathfrak{a}\,|\,\mathfrak{b}) = \sum_{\mu=1}^{m} a_\mu \bar{b}_\mu$$

where $\mathfrak{a} = (a_1, \cdots, a_m) \in \mathbf{C}^m$ and $\mathfrak{b} = (b_1, \cdots, b_m) \in \mathbf{C}^m$. Abbreviate $\mathbf{P}_{m-1} = \mathbf{P}(\mathbf{C}^m)$. If $m = 1$, then $\mathbf{P}_1$ is the complex projective line. Here $\mathbf{P}(\mathbf{C}^1) = \mathbf{P}_1 = \mathbf{C} \cup \{\infty\}$ are identified by

$$\mathbf{P}(1, z) = z \quad \text{if } z \in \mathbf{C},$$

$$\mathbf{P}(0, 1) = \infty.$$

Let $(\mathbf{C}^2)^*$ be the dual space. Then $\mathbf{P}((\mathbf{C}^2)^*) = \mathbf{P}_1 = \mathbf{C} \cup \{\infty\}$ are identified by

$$\mathbf{P}(z, -1) = z \quad \text{if } z \in \mathbf{C},$$

$$\mathbf{P}(1, 0) = \infty.$$

If $a \in \mathbf{P}_1 = \mathbf{P}((\mathbf{C}^2)^*)$ and $z \in \mathbf{P}_1 = \mathbf{P}(\mathbf{C}^2)$, then

$$\|w; a\| = \frac{|w - a|}{\sqrt{1 + |a|^2}\sqrt{1 + |w|^2}} \qquad \text{if } a \neq \infty \neq w,$$

$$\|w; \infty\| = \frac{1}{\sqrt{1 + |w|^2}} \qquad \text{if } w \neq \infty,$$

$$\|\infty; \infty\| = 0.$$

Realize $\mathbf{P}_1$ as a Riemann sphere of diameter 1 in $\mathbf{R}^3$. Then $\|w; a\|$ is the length of the line segment with the edges w, and a.

3. Analytic Chains and Divisors

If A is an analytic subset of an open set in W, let $\mathfrak{R}(A)$ be the set of regular points of A. Now, consider an open subset $G \neq \emptyset$ as an open subset of the hermitian vector space W. A function $\nu : G \to \mathbf{Z}$ is said to be an analytic chain of dimension p on G if and only if

1. the support of ν is an analytic set of pure dimension p or empty,
2. the function $\nu \mid \mathfrak{R}(\operatorname{supp} \nu)$ is locally constant.

Therefore, ν is a nonzero constant on each component of $\mathfrak{R}(\operatorname{supp} \nu)$. Two analytic chains are said to be equivalent if they have the same support A and if they agree pointwise on $\mathfrak{R}(A)$. This defines an equivalence relation. The equivalence classes of analytic chains of dimension p form a module.

Take $0 < R \leqslant +\infty$. Let ν be an analytic chain of dimension p on $W(R)$ with $A = \operatorname{supp} \nu$. The volume function of ν is defined by

$$\sigma_\nu(r) = \int_{A[r]} \nu v^p \quad \text{if } p > 0,$$

$$\sigma_\nu(r) = \sum_{|\mathfrak{z}| \leqslant r} \nu(\mathfrak{z}) \quad \text{if } p = 0.$$

The counting function of ν is defined by

$$n_\nu(r) = \sigma_\nu(r)/r^{2p}.$$

Of course, if $p = 0$, then $n_\nu(r) = \sigma_\nu(r)$.

THEOREM 3.1 (KNESER-LELONG-THIE). *Take* $0 < R \leqslant +\infty$. *Let* ν *be an analytic chain of dimension* $p > 0$. *Define* $A = \operatorname{supp} \nu$. *Let* $f: W(R) \to \mathbf{C}$ *be a holomorphic function. Then*

$$n_\nu(0) = \lim_{r \to 0} n_\nu(r) \in \mathbf{Z}$$

exists where the Lelong number $n_\nu(0)$ *is an integer. If* $0 < r < R$, *then*

$$\frac{1}{r^{2p}} \int_{A(r)} \nu f v^p = \int_{A(r)} \nu f \omega^p + n_\nu(0) f(0).$$

If $f \equiv 1$, *then*

$$n_\nu(r) = \int_{A(r)} \nu \omega^p + n_\nu(0).$$

If $\nu \geqslant 0$, *then* n_ν *increases.*

Historical remarks. Kneser [21] proved the theorem in the case $f \equiv 1$ and $0 \notin \operatorname{supp} \nu$. Lelong [27] showed the theorem for $f \equiv 1$. If $p = m - 1$, the theorem was proved in [46].

Thie [62] showed that the Lelong number is an integer. By parallel translation, the Lelong number of ν can be defined at every point of W. At every point $\mathfrak{z} \in \Re(\operatorname{supp} \nu)$ the Lelong number of ν is given by $\nu(\mathfrak{z})$. If this is true for all $\mathfrak{z} \in W(R)$, then the analytic chain ν is said to be effective. Each equivalence class of analytic chains contains one and only one effective analytic chain. According to King [19], a positive closed current D on $W(R)$ is an effective analytic chain if and only if each Lelong number of D is an integer.

A proof of the full statement of Theorem 3.1 will be given here for the first time. Tung's version of the Stokes theorem [65] will be used and the method resembles Kneser's original method.

PROOF. Since ν is constant on each connectivity component of $\Re(A)$ and since the statement is additive in ν, it suffices to prove the case where $\nu \,|\, \Re(A) \equiv 1$. Observe $dd^c f = 0$ because f is holomorphic. Take $r \in \mathbf{R}[0, R)$. Stokes theorem of Tung [65] implies

$$\int_{A\langle r\rangle} d^c f \wedge \upsilon^{p-1} = \int_{A(r)} dd^c f \wedge \upsilon^{p-1} = 0.$$

Take $s \in \mathbf{R}(0, r)$. Let $g : \mathbf{R}[s^2, r^2] \to \mathbf{C}$ be any function of class C^1. Write $B = A(r) - A[s]$. Let $j : \Re(A) \to W$ be the inclusion map. An easy calculation and the consideration of bidegrees imply

$$j^*(df \wedge d^c g \circ \tau \wedge \upsilon^{p-1}) = j^*(dg \circ \tau \wedge d^c f \wedge \upsilon^{p-1}).$$

Hence

$$\begin{aligned}
&\int_B df \wedge d^c(g \circ \tau) \wedge \upsilon^{p-1} \\
&\quad = \int_B d(g \circ \tau) \wedge d^c f \wedge \upsilon^{p-1} \\
&\quad = \int_B d(g \circ \tau d^c f \wedge \upsilon^{p-1}) \\
&\quad = g(r^2) \int_{A\langle r\rangle} d^c f \wedge \upsilon^{p-1} - g(s^2) \int_{A\langle s\rangle} d^c f \wedge \upsilon^{p-1} \\
&\quad = 0.
\end{aligned}$$

At first, take $g(t) = t$ and let s converge to zero. Then

$$\int_{A(r)} df \wedge d^c \tau \wedge \upsilon^{p-1} = 0$$

results. Therefore

$$\begin{aligned}
\int_{A(r)} f \upsilon^p &= \frac{1}{4\pi} \int_{A(r)} f dd^c \tau \wedge \upsilon^{p-1} \\
&= \frac{1}{4\pi} \int_{A(r)} d(f d^c \tau \wedge \upsilon^{p-1}) \\
&= \frac{1}{4\pi} \int_{A\langle r\rangle} f d^c \tau \wedge \upsilon^{p-1}.
\end{aligned}$$

Now, another choice of g will be made. If $p = 1$, take $g(t) = \log t$. If $p > 1$, take $g(t) = -\,t^{1-p}/(p-1)$. Then $d^c g \circ \tau = \tau^{-p} d^c \tau$ and

$$d^c(g \circ \tau) \wedge \upsilon^{p-1} = \tau^{-p} d^c \tau \wedge \upsilon^{p-1} = d^c \log \tau \wedge \omega^{p-1}.$$

Therefore

$$\int_B df \wedge d^c \log \tau \wedge \omega^{p-1} = \int_B df \wedge d^c g \circ \tau \wedge \upsilon^{p-1} = 0.$$

Hence

$$\begin{aligned}\int_B f\omega^p &= \frac{1}{4\pi}\int_B fdd^c \log \tau \wedge \omega^{p-1} \\ &= \frac{1}{4\pi}\int_B d(fd^c \log \tau \wedge \omega^{p-1}) \\ &= \frac{1}{4\pi}\int_B d(f\tau^{-p} d^c \tau \wedge \upsilon^{p-1}) \\ &= \frac{1}{4\pi}\frac{1}{r^{2p}}\int_{A\langle r\rangle} fd^c\tau \wedge \upsilon^{p-1} - \frac{1}{4\pi}\frac{1}{s^{2p}}\int_{A\langle s\rangle} fd^c\tau \wedge \upsilon^{p-1} \\ &= \frac{1}{r^{2p}}\int_{A(r)} f\upsilon^p - \frac{1}{s^{2p}}\int_{A(s)} f\upsilon^p.\end{aligned}$$

If $f \equiv 1$, this implies

$$\begin{aligned}0 \leqslant \int_B \omega^p &= \frac{1}{r^{2p}}\int_{A(r)} \upsilon^p - \frac{1}{s^{2p}}\int_{A(s)} \upsilon^p \\ &= n_\nu(r) - n_\nu(s).\end{aligned}$$

Hence the function n_ν increases. Therefore, the limit $n_\nu(s) \to n_\nu(0)$ for $s \to 0$ exists.

Now, return to the case of an arbitrary holomorphic function f on $W(R)$. Take any $\epsilon > 0$. A number $s(\epsilon)$ with $0 < s(\epsilon) < r$ exists such that $|f(\mathfrak{z}) - f(0)| < \epsilon$ for all $\mathfrak{z} \in W[s(\epsilon)]$ and $|n_\nu(s) - n_\nu(0)| < \epsilon$ for all $0 < s < s(\epsilon)$. Then

$$\left|\frac{1}{s^{2p}}\int_{A(s)} f\upsilon^p - f(0)n_\nu(0)\right| \leqslant \epsilon(n_\nu(0) + \epsilon) + \epsilon|f(0)|.$$

Therefore

$$\frac{1}{s^{2p}}\int_{A(s)} f\upsilon^p \longrightarrow f(0)n_\nu(0) \quad \text{for } s \longrightarrow 0.$$

Hence

$$\int_{A(r)} f\omega^p = \frac{1}{r^{2p}}\int_{A(r)} f\upsilon^p - f(0)n_\nu(0). \quad \text{q. e. d.}$$

LEMMA 3.2. *Take* $0 < R \leqslant +\infty$. *Let* ν *be an analytic chain of dimension* $p > 0$ *on* $W(R)$. *Define* $A = \operatorname{supp} \nu$. *Take* $0 \leqslant s < r < R$. *Let* $f: \mathbf{R}[s, r] \to \mathbf{C}$ *be the function of class* C^1. *Define* $B = A(r) - A[s]$. *Then*

$$\int_D \nu f \circ \tau_0 \upsilon^p = \sigma_\nu(r)f(r) - \sigma_\nu(s)f(s) - \int_s^r \sigma_\nu(t)f'(t)\,dt,$$

$$\int_D \nu f \circ \tau_0 \omega^p = n_\nu(r)f(r) - n_\nu(s)f(s) - \int_0^r n_\nu(t)f'(t)\,dt.$$

(The proof is easily obtained along the lines of the proof of Lemma 2.1 in [59].)

LEMMA 3.3. *Take* $0 < R \leqslant +\infty$. *Let* $\nu \geqslant 0$ *be an analytic set of dimension* $p > 0$ *on* $W(R)$. *Define* $A = \operatorname{supp} \nu$. *Take* $0 \leqslant s < R$. *Define* $B = A - A(s)$. *Let* $f: \mathbf{R}[s, R) \to \mathbf{R}_+$ *be a decreasing function of class* C^1. *Assume* $f(x) \to 0$ *for* $x \to R$. *If either one of the following two integrals exist, then both exist and*

$$\int_B \nu f \circ \tau_0 \upsilon^p = -\sigma_\nu(s) f(s) - \int_s^r \sigma_\nu(t) f'(t)\, dt.$$

Moreover $\sigma_\nu(r) f(r) \to 0$ *for* $r \to R$. *Similarly, if either one of the following two integrals exist, then both exist and*

$$\int_B \nu f \circ \tau_0 \omega^p = -n_\nu(s) f(s) - \int_s^r \sigma_\nu(t) f'(t)\, dt.$$

Moreover, $n_\nu(r) f(r) \to 0$ *for* $r \to R$.

(The proof is easily obtained along the lines of the proof of Lemma 5.6 in [53].)

Take $0 < R \leqslant +\infty$. Let ν be an analytic chain on $W(R)$ of dimension p. For $0 < s < r < R$ define the valence function of ν by

$$N_\nu(r, s) = \int_s^r n_\nu(t) \frac{dt}{t}.$$

If $0 \notin \operatorname{supp} \nu$, then $s = 0$ is permitted;

$$N_\nu(r) = N_\nu(r, 0) = \int_0^r n_\nu(t) \frac{dt}{t}.$$

Take $0 < s < r < +\infty$ and $p > 0$. Define

$$\phi_{sr}(x) = \begin{cases} \log^+ r/x & \text{if } x > s, \\ \log r/s & \text{if } 0 \leqslant x \leqslant s. \end{cases}$$

$$\phi_{0r}(x) = \log^+ r/x \quad \text{if } x > 0.$$

$$\psi_{sr}(x) = \begin{cases} \dfrac{1}{2p}\,[x^{-2p} - r^{-2p}] & \text{if } s < x \leqslant r, \\[2ex] \dfrac{1}{2p}\,[s^{-2p} - r^{-2p}] & \text{if } 0 \leqslant x \leqslant s. \end{cases}$$

$$\psi_{0r}(x) = \frac{1}{2p}\,[x^{-2p} - r^{-2p}] \quad \text{if } 0 < x \leqslant r.$$

Assume $0 < s < r < R$ and $p > 0$. The last two lemmata imply

$$N_\nu(r, s) = \int_{A(r)} \nu \phi_{sr} \circ \tau_0 \omega^p + \phi_{sr}(0) n_\nu(0),$$

$$N_\nu(r, s) = \int_{A(r)} \nu \psi_{sr} \circ \tau_0 \upsilon^p.$$

If $0 \notin \operatorname{supp} \nu$, then

$$N_\nu(r) = \int_{A(r)} \nu \phi_{0r} \circ \tau_0 \omega^p = \int_{A(r)} \nu \psi_{0r} \circ \tau_0 \upsilon^p.$$

If $p = 0$ and $0 < s < r < R$, then

$$N_\nu(r, s) = \sum_{\mathfrak{z} \in A(r)} \nu(\mathfrak{z}) \phi_{sr}(|\mathfrak{z}|).$$

If $p = 0$ and $0 \notin \operatorname{supp} \nu$, then

$$N_\nu(r) = \sum_{\mathfrak{z} \in A(r)} \nu(\mathfrak{z}) \phi_{0r}(|\mathfrak{z}|).$$

The following theorem is essentially due to Shiffman [40]. The proof can be obtained by fiber integration. Since the theorem will not be applied here, no proof shall be given here.

THEOREM 3.4. CROFTON'S FORMULA (SHIFFMAN). *Take* $0 < R \leqslant +\infty$. *Let* ν *be an analytic chain of dimension* p *on* $W(R)$ *with* $0 < p < m$. *Define* $q = m - p - 1$ *and* $A = \operatorname{supp} \nu$. *Assume that* $\dim A \cap E(a) \leqslant 0$ *for almost all* $a \in G_q(W)$. *Let* $F : W(R) \to \mathbf{C}$ *be a function. Assume that* $\nu F \omega^p$ *is integrable over* A. *Then*

$$H(a) = \sum_{A \cap E(a)} \nu F$$

converges absolutely for almost all $a \in G_q(W)$ *and is integrable over the Grassmann manifold* $G_q(W)$. *Moreover*

$$\int_A \nu F \omega^p = \frac{1}{\Delta(m, q)} \int_{G_q(W)} H \ddot{\omega}_q^{d(m,q)}.$$

Now, the special case of a divisor shall be discussed. Let $G \neq \varnothing$ be an open subset in W. Let f be a holomorphic function on G. Take $\mathfrak{a} \in G$. Let $G_{\mathfrak{a}}$ be the connectivity component of G containing $\mathfrak{a}$. Assume $f | G_{\mathfrak{a}} \not\equiv 0$. Then a series

$$f(\mathfrak{z}) = \sum_{\lambda = p}^{\infty} P_\lambda(\mathfrak{z} - \mathfrak{a})$$

converges on some neighborhood of $\mathfrak{a}$ and represents f on this neighborhood. Here P_λ is a homogeneous polynomial of degree λ and $P_p \not\equiv 0$. The polynomials P_λ depend on f and $\mathfrak{a}$ only. The number

$$\mu_f(\mathfrak{a}) = \mu_f^0(\mathfrak{a}) = p$$

is called the zero multiplicity of f at $\mathfrak{a}$.

Let f be a meromorphic function on G. Take $\mathfrak{a} \in G$ and $c \in \mathbf{P}_1$. Again let $G_{\mathfrak{a}}$ be the component of G containing $\mathfrak{a}$. Assume that $f | G_{\mathfrak{a}} \not\equiv c$. Then an open, connected neighborhood U of $\mathfrak{a}$ in G and holomorphic functions $g \not\equiv 0$ and $h \not\equiv 0$ exist on U such that $h \cdot f | U = g$ and $\dim g^{-1}(0) \cap h^{-1}(0) \leqslant m - 2$. Then

$$\mu_f^c(\mathfrak{a}) = \mu_{g - ch}^0(\mathfrak{a}) \quad \text{if } c \in \mathbf{C},$$

$$\mu_f^\infty(\mathfrak{a}) = \mu_h^0(\mathfrak{a}) \qquad \text{if } c = \infty$$

are well defined and called the c-multiplicity of f. The function $\mu_f^c : G \to \mathbf{Z}$ is a nonnegative, effective analytic chain of dimension $m - 1$ on G, called the c-divisor of f. If $f \not\equiv 0$ on each component on G, then

$$\mu_f = \mu_f^0 - \mu_f^\infty$$

is called the divisor of f. The function μ_f is an effective, analytic chain of dimension $m - 1$ on G. The function f is holomorphic on G, if and only if $\mu_f \geqslant 0$. If f is meromorphic on G, and if f is considered as a meromorphic map into $\mathbf{P}_1$, then

$$f^{-1}(c) = \operatorname{supp} \mu_f^c.$$

If $f_1 \not\equiv 0$ and $f_2 \not\equiv 0$ are meromorphic on G, then

$$\mu_{f_1 \cdot f_2} = \mu_{f_1} + \mu_{f_2},$$

$$\mu_{f_1+f_2} \geqslant \text{Max}\,(\mu_{f_1}, \mu_{f_2}) \quad \text{if } f_1 + f_2 \not\equiv 0.$$

A function $\nu : G \to \mathbf{Z}$ is said to be a divisor if and only if for each $\mathfrak{a} \in G$ an open, connected neighborhood U of $\mathfrak{a}$ in G and a meromorphic function $f \not\equiv 0$ exist such that $\nu \mid U = \mu_f$. The divisors on G form a module. The divisors are precisely the effective analytic chains of dimension $m-1$. If f is a meromorphic function on G, the functions μ_f and μ_f^c are divisors if defined.

A divisor $\nu : G \to \mathbf{Z}$ is nonnegative, if and only if for each $\mathfrak{a} \in G$ an open, connected neighborhood U of $\mathfrak{a}$ in G and a holomorphic function $f \not\equiv 0$ on U exist such that $\nu \mid U = \mu_f$.

A divisor $\nu : G \to \mathbf{Z}$ is said to be a principal divisor, if and only if $\nu = \mu_f$ for some meromorphic function f on G. The open set G is called a Cousin-II-domain if and only if each divisor on G is a principal divisor. If $0 < R \leqslant +\infty$, then $W(R)$ is a Cousin-II-domain. Let ν be a divisor on $W(R)$. Then ν is an effective, analytic chain of dimension $m-1$ on $W(R)$. Hence the counting function n_ν and the valence function N_ν are defined. If f is a meromorphic function on $W(R)$, and if $\nu = \mu_f$, write

$$N_f = N_\nu, \qquad n_\nu = n_f.$$

Let W and V be hermitian vector spaces. Let $G \neq \varnothing$ and $H \neq \varnothing$ be open, connected subsets of W and V respectively. Let $\phi : H \to G$ be a holomorphic map. Let ν be a divisor on G such that $\phi(H) \not\subseteq \text{supp}\,\nu$. Then one and only one divisor $\phi^*(\nu)$ exists on H satisfying the following condition.

Let G_0 be open and connected in G with $H_0 = \phi^{-1}(G_0) \neq \varnothing$. Assume that $g \not\equiv 0$ and $h \not\equiv 0$ are holomorphic on G_0 such that $\nu \mid G_0 = \mu_g - \mu_h$. Then

$$\phi^*(\nu) \mid H_0 = \mu_{g \circ \phi} - \mu_{h \circ \phi}.$$

Take $0 < R \leqslant +\infty$. Let ν be a divisor on $W(R)$. Take a vector $\mathfrak{a} \neq 0$ in W. Define $j_\mathfrak{a} : \mathbf{C} \to W$ by $j_\mathfrak{a}(z) = z\mathfrak{a}$. Then $j_\mathfrak{a} : \mathbf{C}(R/|\mathfrak{a}|) \to W(R)$ is holomorphic. Also $\text{Im}\, j_\mathfrak{a} \not\subseteq \text{supp}\,\nu$ for almost all $\mathfrak{a}$. If so, then $\nu[\mathfrak{a}] = j_\mathfrak{a}^*(\nu)$ is a divisor on $\mathbf{C}(R/|\mathfrak{a}|)$. Also write $\nu[\mathfrak{a}]\,(z) = \nu(z;\mathfrak{a})$, and

$$n_{\nu[\mathfrak{a}]}(r) = n_\nu(r;\mathfrak{a}), \qquad N_{\nu[\mathfrak{a}]}(r, s) = N_\nu(r, s; \mathfrak{a}).$$

Now, the Crofton formula for divisors reads as follows [46]:

THEOREM 3.5. CROFTON'S FORMULA FOR DIVISORS. *Take* $0 < R \leqslant +\infty$. *Let* ν *be a divisor on* $W(R)$. *Define* $A = \text{supp}\,\nu$. *Let* $F : W(R) \to \mathbf{C}$ *be a function. Assume that* $\nu F \omega^{m-1}$ *is integrable over* A. *Then*

$$J(\mathfrak{a}) = \sum_{0 < |z| < R} \nu(z;\mathfrak{a}) F(z\mathfrak{a})$$

converges for almost all $\mathfrak{a} \in W\langle 1\rangle$. *If* $\phi \in \mathbf{R}$, *then*

$$J(e^{i\phi}\mathfrak{a}) = J(\mathfrak{a}).$$

A function $\ddot{J}$ *exists almost everywhere on* $\mathbf{P}(W)$ *such that* $\ddot{J} \circ P = J$ *almost everywhere on* $W\langle 1\rangle$. *Moreover,*

$$\int_A F\omega^{m-1} = \int_{W\langle 1\rangle} J\sigma = \int_{\mathbf{P}(W)} \ddot{J}\ddot{\omega}^{m-1}.$$

Crofton's formula implies

$$n_\nu(r) = \int_{W\langle 1\rangle} n_\nu(r;\mathfrak{a})\,\sigma(\mathfrak{a}), \quad N_\nu(r, s) = \int_{W\langle 1\rangle} N_\nu(r, s;\mathfrak{a})\,\sigma(\mathfrak{a}),$$

if $0 < s < r < R$. If $0 \notin \operatorname{supp} \nu$, then

$$N_\nu(r) = \int_{W\langle 1\rangle} N_\nu(r, \mathfrak{a})\,\sigma(\mathfrak{a})$$

for $0 < r < R$. These identities are due to Kneser [21]. They imply immediately the Jensen formula.

Theorem 3.6. Jensen formula [49]. *Take $0 < R \leqslant +\infty$. Let $f \not\equiv 0$ be a meromorphic function on $W(R)$. Let $0 < s < r < R$. Then*

$$N_f(r, s) = \int_{W\langle r\rangle} \log|f|\,\sigma - \int_{W\langle s\rangle} \log|f|\,\sigma.$$

If f is holomorphic at $0 \in W$ and if $f(0) \neq 0$, then

$$N_f(r) = \int_{W\langle r\rangle} \log|f|\,\sigma - \log|f(0)|.$$

Proof. Define $\nu = \mu_f$. Then

$$\begin{aligned} N_f(r, s) &= \int_{W\langle 1\rangle} N_\nu(r, s;\mathfrak{a})\,\sigma(\mathfrak{a}) \\ &= \int_{W\langle 1\rangle} \frac{1}{2\pi}\int_0^{2\pi} \log|f(re^{i\phi}\mathfrak{a})|\,d\phi\,\sigma(\mathfrak{a}) \\ &\quad - \int_{W\langle 1\rangle} \frac{1}{2\pi}\int_0^{2\pi} \log|f(se^{i\phi}\mathfrak{a})|\,d\phi\,\sigma(\mathfrak{a}) \\ &= \int_{W\langle r\rangle} \log|f|\,\sigma_r - \int_{W\langle s\rangle} \log|f|\,\sigma_s. \end{aligned}$$

The second formula is obtained by the limit $s \to 0$. q.e.d.

Let $G \neq \varnothing$ be an open subset of the vector space W. A function $u : G \to \mathbf{R} \cup \{-\infty\}$ is said to be pluri-subharmonic if and only if the following conditions hold:

(1) If G_0 is a connectivity component of G, then $U\,|\,G_0 \not\equiv -\infty$.

(2) The function u is upper semicontinuous.

(3) Let $\mathfrak{a} \in W$ and $0 \neq \mathfrak{b} \in W$. Define $L = \mathfrak{a} + \mathbf{C}\mathfrak{b}$ and assume that $G \cap L \neq \varnothing$. Assume that $u\,|\,G \cap L \not\equiv -\infty$ on each connectivity component of $G \cap L$. Then $u\,|\,G \cap L$ is subharmonic.

If $u : G \to \mathbf{R}$ is a pluri-subharmonic function of class C^2, then $dd^c u \geqslant 0$. A function $u : G \to \mathbf{R}$ of class C^2 is said to be pluri-harmonic if and only if $dd^c u \equiv 0$, or, equivaalently if u is locally the real part of a holomorphic function.

Theorem 3.7 (Kneser [21]). *Take $0 < r < R \leqslant +\infty$. Let u be a pluri-subharmonic function on $W(R)$. Define $u^+ = \operatorname{Max}(0, u)$. Let θ be a number with $0 < \theta < 1$. If $\mathfrak{z} \in W(r)$, then*

$$u(\mathfrak{z}) \leqslant r^{2m-2} \int_{W\langle r\rangle} u(\mathfrak{y}) \frac{r^2 - |\mathfrak{z}|^2}{|\mathfrak{y} - \mathfrak{z}|^{2m}} \sigma(\mathfrak{y}).$$

If u *is pluri-harmonic, equality holds. If* $\mathfrak{z} \in W(\theta r)$, *then*

$$u(\mathfrak{z}) \leqslant \frac{1+\theta}{(1-\theta)^{2m-1}} \int_{W\langle r\rangle} u^+ \sigma.$$

Jensen's formula and Theorem 3.7 immediately imply

PROPOSITION 3.8 (RONKIN [35], KUJALA [24]). *Take* $0 < R \leqslant +\infty$. *Let* $\nu \geqslant 0$ *be a divisor on* $W(R)$ *with* $0 \notin \operatorname{supp} \nu$. *Then*

(1) *The function* $N_\nu(r;\mathfrak{a})$ *is continuous in* $(r;\mathfrak{a})$ *for* $|r\,\mathfrak{a}| < R$.

(2) *For each fixed number* $r \in \mathbf{R}(0, R)$, *the function* $N_\nu(r;\mathfrak{a})$ *of* $\mathfrak{a}$ *is pluri-subharmonic for* $|\mathfrak{a}| < R/r$.

(3) *If* $0 < \theta < 1$ *and* $\mathfrak{a} \in W[1]$ *and* $0 < r < \theta R$, *then*

$$N_\nu(r;\mathfrak{a}) \leqslant \frac{1+\theta}{(1-\theta)^{2m-1}} N_\nu\left(\frac{r}{\theta}\right).$$

PROOF. A holomorphic function $f \not\equiv 0$ exists on $W(R)$ with $f(0) = 1$ and $\mu_f = \nu$. Then

$$N_\nu(r;\mathfrak{a}) = \frac{1}{2\pi} \int_0^{2\pi} \log |f(re^{i\phi}\mathfrak{a})|\, d\phi$$

is continuous in r and $\mathfrak{a}$ and pluri-subharmonic in $\mathfrak{a}$ for each fixed r if $|\mathfrak{a}| < R/r$. Theorem 3.7 implies

$$N_\nu(r;\mathfrak{a}) \leqslant \frac{1+\theta}{(1-\theta)^{2m-1}} N_\nu(r)$$

for $|\mathfrak{a}| \leqslant \theta$. If $|\mathfrak{a}| \leqslant 1$ and $r < \theta R$, then

$$N_\nu(r;\mathfrak{a}) \leqslant N_\nu\left(\frac{r}{\theta}; \theta\mathfrak{a}\right) \leqslant \frac{1+\theta}{(1-\theta)^{2m-1}} N_\nu\left(\frac{r}{\theta}\right). \quad \text{q.e.d.}$$

4. The Two Main Theorems

Let V and W be hermitian vector spaces of $\dim V = n+1 > 0$ and $\dim W = m > 0$. Take $0 < R \leqslant +\infty$. Let $f: W(R) \to \mathbf{P}(V)$ be a meromorphic map. Then there exists a largest open subset G of $W(R)$ such that $I_f = W(R) - G$ is analytic with $\dim I_f \leqslant m-2$ and such that $f: G \to \mathbf{P}(V)$ is a holomorphic map. A holomorphic vector function $\mathfrak{v}: W(R) \to V$ exists such $\mathfrak{v}^{-1}(0) = I_f$ and $\mathbf{P}(\mathfrak{v}(\mathfrak{z})) = f(\mathfrak{z})$. For the purposes here, this can suffice as a definition of a meromorphic map $f: W(R) \rightharpoonup \mathbf{P}(V)$. If $\tilde{\mathfrak{v}}$ is another such vector function, then there exists one and only one holomorphic function $h: W(R) \to \mathbf{C} - \{0\}$ such that $\tilde{\mathfrak{v}} = h\mathfrak{v}$.

If $a \in \mathbf{P}(V^*)$, then

$$f^{-1}(\ddot{E}[a]) = \mathfrak{v}^{-1}(E[a]).$$

Assume that f does not map into any proper projective subspace of $\mathbf{P}(V)$, that is

$$f^{-1}(\ddot{E}[a]) \neq W(R) \quad \text{for all } a \in \mathbf{P}(V^*).$$

Take any $\alpha \in \mathbf{P}^{-1}(a)$. Then $\alpha \circ \mathfrak{v} \not\equiv 0$ is a holomorphic function on $W(R)$. The divisor $\mu_f^a = \mu_{\alpha \circ \mathfrak{v}} \geqslant 0$ does not depend of the choices of α and $\mathfrak{v}$ but on a and f only. The divisor μ_f^a is called the a-divisor of f. Observe

$$\operatorname{supp} \mu_f^a = f^{-1}(\ddot{E}[a]).$$

If $\nu = \mu_f^a$, write

$$N_f^a = N_\nu, \qquad n_f^a = n_\nu.$$

If $a \in \mathbf{P}(V^*)$ and $b \in \mathbf{P}(V^*)$, take $\alpha \in \mathbf{P}^{-1}(a)$ and $\beta \in \mathbf{P}^{-1}(b)$ with $|\alpha| = 1 = |\beta|$. Jensen's formula implies

$$N_f^a(r,s) - N_f^b(r,s) = \int_{W\langle r\rangle} \log\left|\frac{\alpha \circ \mathfrak{v}}{\beta \circ \mathfrak{v}}\right| \sigma - \int_{W\langle s\rangle} \log\left|\frac{\alpha \circ \mathfrak{v}}{\beta \circ \mathfrak{v}}\right| \sigma$$

for $0 < s < r < R$. Here

$$\log\left|\frac{\alpha \circ \mathfrak{v}}{\beta \circ \mathfrak{v}}\right| = \log\frac{|\alpha \circ \mathfrak{v}|}{|\mathfrak{v}|} - \log\frac{|\beta \circ \mathfrak{v}|}{|\mathfrak{v}|}$$
$$= \log \|f, a\| - \log \|f, b\|.$$

For $0 < r < R$, define the compensation function of f for a by

$$m_f^a(r) = \int_{W\langle r\rangle} \log\frac{1}{\|f, a\|}\, \sigma \geqslant 0.$$

The separation of terms in a and b yields

$$N_f^a(r, s) + m_f^a(r) - m_f^a(s) = N_f^b(r, s) + m_f^b(r) - m_f^b(s).$$

Hence, the characteristic

$$T_f(r, s) = N_f^a(r, s) + m_f^a(r) - m_f^a(s) \tag{4.1}$$

of f does not depend on $a \in \mathbf{P}(V^*)$ for $0 < s < r < R$. The First Main Theorem (4.1) is established. Assume that f is holomorphic at 0. For all $a \in \mathbf{P}(V^*)$ with $a \neq f(0)$ define

$$m_f^a(0) = \log \frac{1}{\| f(0), a \|} \geqslant 0.$$

Then the characteristic $T_f(r) = N_f^a(r) + m_f^a(r) - m_f^a(0)$ of f does not depend on $a \in \mathbf{P}(V^*) - \{f(0)\}$ if $0 < r < R$.

Return to the case where f may or may not be holomorphic at $0 \in W$. An easy computation shows

$$\int_{\mathbf{P}(V^*)} \log \frac{1}{\| w, a \|} \ddot{\omega}^n(a) = \frac{1}{2} \sum_{\nu=1}^{n} \frac{1}{\nu}$$

which implies by an exchange of integrals

$$\int_{\mathbf{P}(V^*)} m_f^a(r) \ddot{\omega}^n(a) = \frac{1}{2} \sum_{\nu=1}^{n} \frac{1}{\nu}.$$

The integration of the First Main Theorem yields

$$T_f(r, s) = \int_{\mathbf{P}(V^*)} N_f^a(r, s) \ddot{\omega}^n(a)$$

if $0 < s < r < R$. If f is holomorphic at 0, the same identity holds with $T_f(r, 0) = T_f(r)$ and $N_f^a(r, 0) = N_f^a(r)$.

By an exchange of integration, the following explicit representation of the characteristic is obtained:

$$A_f(t) = \frac{1}{t^{2m}} \int_{W(t)} f^*(\ddot{\omega}_V) \wedge v_W^{m-1} \geqslant 0,$$

$$T_f(r, s) = \int_s^r A_f(t) \frac{dt}{t} \geqslant 0.$$

It can be shown that A_f is an increasing function. Hence

$$A_f(0) = \lim_{t \to 0} A_f(t)$$

exists. It can be shown that $A_f(0) \geqslant 0$ is an integer, and that

$$A_f(t) = \int_{W(t)} f^*(\ddot{\omega}_V) \wedge \omega_W^{m-1} + A_f(0).$$

The function A_f is continuous. The function T_f increases in r and decreases in s and is of class C^1. If f is holomorphic at $0 \in W$, then $s = 0$ is permitted;

$$T_f(r) = \int_0^r A_f(t) \frac{dt}{t}.$$

Let $\mathfrak{w} : W(R) \to V$ be a holomorphic map such that $\mathfrak{w} \not\equiv 0$ and such that $f(\mathfrak{z}) = \mathbf{P}(\mathfrak{w}(\mathfrak{z}))$ if $\mathfrak{z} \in W(R) - I_f$ and $\mathfrak{w}(\mathfrak{z}) \neq 0$. Then $\mathfrak{w}^{-1}(0) \supseteq I_f$. Take any holomorphic map $\mathfrak{v} : W(R) \to V$ such that $\mathfrak{v}^{-1}(0) = I_f$ and $\mathbf{P} \circ \mathfrak{v} = f$ on $W(R) - I_f$. Then one and only one holomorphic function h exists on $W(R)$ such that $\mathfrak{w} = h\mathfrak{v}$. Then the nonnegative divisor $d_{\mathfrak{w}} = \mu_h$ depends on $\mathfrak{w}$ only and not on the choice of $\mathfrak{v}$. Here $d_{\mathfrak{w}}$ is the greatest common divisor of $\mathfrak{w}$.

Take $a \in \mathbf{P}(V^*)$ and $\alpha \in \mathbf{P}^{-1}(a)$ with $|\alpha| = 1$. Then $\alpha \circ \mathfrak{w} = h \cdot (\alpha \circ \mathfrak{v})$. Therefore $\mu_{\alpha \circ \mathfrak{w}} = d_{\mathfrak{w}} + \mu_f^a$. Take $0 < s < r < R$. Then

$$N_f^a(r, s) = N_{\alpha \circ \mathfrak{w}}(r, s) - N_{d_{\mathfrak{w}}}(r, s).$$

Jensen's formula implies

$$N_{\alpha \circ \mathfrak{w}}(r, s) = \int_{W\langle r\rangle} \log|\alpha \circ \mathfrak{v}|\,\sigma - \int_{W\langle s\rangle} \log|\alpha \circ \mathfrak{w}|\,\sigma.$$

By definition

$$m_f^a(r) = \int_{W\langle r\rangle} \log \frac{|\mathfrak{v}|}{|\alpha \circ \mathfrak{v}|}\,\sigma = \int_{W\langle r\rangle} \log \frac{|\mathfrak{w}|}{|\alpha \circ \mathfrak{w}|}\,\sigma.$$

Hence the First Main Theorem implies

$$T_f(r, s) = \int_{W\langle r\rangle} \log|\mathfrak{w}|\,\sigma - \int_{W\langle s\rangle} \log|\mathfrak{w}|\,\sigma - N_{d_{\mathfrak{w}}}(r, s).$$

Because $N_{d_{\mathfrak{w}}}(r, s) \geqslant 0$, this implies

$$T_f(r, s) \leqslant \int_{W\langle r\rangle} \log|\mathfrak{w}|\,\sigma - \int_{W\langle s\rangle} \log|\mathfrak{w}|\,\sigma.$$

If $\dim \mathfrak{w}^{-1}(0) \leqslant m - 2$, then $d_{\mathfrak{w}} \equiv 0$ and

$$T_f(r, s) = \int_{W\langle r\rangle} \log|\mathfrak{w}|\,\sigma - \int_{W\langle s\rangle} \log|\mathfrak{w}|\,\sigma.$$

If $\mathfrak{w}(0) \neq 0$, then limit $s \to 0$ implies

$$\begin{aligned} T_f(r) &= \int_{W\langle r\rangle} \log|\mathfrak{w}|\,\sigma - \log|\mathfrak{w}(0)| - N_{d_{\mathfrak{w}}}(r) \\ &\leqslant \int_{W\langle r\rangle} \log|\mathfrak{w}|\,\sigma - \log|\mathfrak{w}(0)|. \end{aligned}$$

If $\dim \mathfrak{w}^{-1}(0) \leqslant m - 2$ and $\mathfrak{w}(0) \neq 0$, then

$$T_f(r) = \int_{W\langle r\rangle} \log|\mathfrak{w}|\,\sigma - \log|\mathfrak{w}(0)|.$$

Now, assume that the meromorphic map $f\colon W(R) \to \mathbf{P}(V)$ is holomorphic at $0 \in W$. Take any vector $\mathfrak{a} \neq 0$ in W. Define

$$j_{\mathfrak{a}} : \mathbf{C}(R/|\mathfrak{a}|) \longrightarrow W(R)$$

by $j_{\mathfrak{a}}(z) = z\mathfrak{a}$. Then

$$f \circ j_{\mathfrak{a}} : \mathbf{C}(R/|\mathfrak{a}|) \longrightarrow \mathbf{P}(V)$$

is a holomorphic map. Let $T_f(r, s; \mathfrak{a})$, $N_f^a(r, s; \mathfrak{a})$, $m_f^a(r; \mathfrak{a})$ be the characteristic, valence, compensation functions respectively of $f \circ j_{\mathfrak{a}}$. Then

$$\begin{aligned} N_f^a(r, s) &= \int_{W\langle 1\rangle} N_f^a(r, s; \mathfrak{a})\,\sigma(\mathfrak{a}), \\ m_f^a(r) &= \int_{W\langle 1\rangle} m_f^a(r; \mathfrak{a})\,\sigma(\mathfrak{a}). \end{aligned}$$

The First Main Theorem implies

$$T_f(r, s) = \int_{W\langle 1\rangle} T_f(r, s; \mathfrak{a})\sigma(\mathfrak{a})$$

for $0 \leqslant s < r < R$.

Let $\mathfrak{v} : W(R) \to V$ be a holomorphic map with $I_f = \mathfrak{v}^{-1}(0)$ and $f = \mathbf{P} \circ \mathfrak{v}$ on $W(R) - I_f$. Then $\mathfrak{v}(0) \neq 0$. Let S be the set of all $\mathfrak{a} \in W$ such that $\mathfrak{v}(z\mathfrak{a}) = 0$ for some $z \in \mathbf{C}$ with $|z\mathfrak{a}| < R$. Then S has measure zero on W. Take $0 < r < R$. Take $0 \neq \mathfrak{a} \in W(r^{-1}R)$. Then

$$T_f(r; \mathfrak{a}) \leqslant \frac{1}{2\pi}\int_0^{2\pi} \log|\mathfrak{v}(re^{i\phi}\mathfrak{a})|\,d\phi - \log|\mathfrak{v}(0)|.$$

If $\mathfrak{a} \notin S$, then

$$T_f(r; \mathfrak{a}) = \frac{1}{2\pi}\int_0^{2\pi} \log|\mathfrak{v}(re^{i\phi}\mathfrak{a})|\,d\phi - \log|\mathfrak{v}(0)|.$$

In any event, a nonnegative pluri-subharmonic function V_r on $W(r^{-1}R)$ is defined by

$$V_r(\mathfrak{a}) = \frac{1}{2\pi}\int_0^{2\pi} \log|\mathfrak{v}(re^{i\phi}\mathfrak{a})|\,d\phi - \log|\mathfrak{v}(0)|$$

for $|\mathfrak{a}| < r^{-1}R$ such that $T_f(r) = \int_{W\langle 1\rangle} V_r\sigma$. Observe that $W\langle 1\rangle \subseteq W(r^{-1}R)$. Hence the integral is defined. Take θ with $0 < \theta < 1$ and $r < \theta R$. Theorem 3.7 implies

$$T_f(r; \mathfrak{a}) \leqslant \frac{1+\theta}{(1-\theta)^{2m-1}} T_f\left(\frac{r}{\theta}\right) \tag{4.2}$$

for all $0 \neq \mathfrak{a} \in W[1]$. Let H be a linear subspace of dimension p of W with $0 < p < m$. Integration over $H\langle 1\rangle$ implies

$$T_{f|H}(r) \leqslant \frac{1+\theta}{(1-\theta)^{2m-1}} T_f\left(\frac{r}{\theta}\right) \tag{4.3}$$

if $0 < \theta < 1$ and $0 < r < \theta R$ and if f is holomorphic at $0 \in W$.

Take $R = +\infty$. Let $f : W \to \mathbf{P}(V)$ be a meromorphic map. Assume that $f(W) \not\subseteq \ddot{E}[a]$ for each $a \in \mathbf{P}(V^*)$. Then

$$T_f(r, s) \longrightarrow \infty \quad \text{for } r \longrightarrow \infty.$$

The defect

$$0 \leqslant \delta_f(a) = \liminf_{r\to\infty} \frac{m_f^a(r)}{T_f(r, s)} = 1 - \limsup_{r\to\infty} \frac{N_f^a(r, s)}{T_f(r, s)} \leqslant 1$$

of f is defined for all $a \in \mathbf{P}(V^*)$ and is independent of s. If $f(W) \cap \ddot{E}[a] = \varnothing$, then $\delta_f(a) = 1$. The meromorphic map f is said to be rational if and only if there exists a holomorphic vector function $\mathfrak{v} : W \to V$ with polynomial coordinate functions such that $\mathfrak{v}^{-1}(0) = I_f$ and $\mathbf{P} \circ \mathfrak{v} = f$ on $W - I_f$. The meromorphic map f is not rational if and only if

$$T_f(r, s)/\log r \longrightarrow \infty \quad \text{for } r \longrightarrow \infty$$

for some s, hence for all $s > 0$. (See [49, Satz 24.1].) A nonnegative divisor $\nu : W \to \mathbf{Z}_+$ is said to be algebraic if and only if ν is the zero divisor of a polynomial. According to Rutishauser [39] (see also [50, p. 295]), a divisor $\nu : W \to \mathbf{Z}_+$ is algebraic if and only if

n_ν is bounded, which implies $N_\nu = O(\log r)$. Therefore $\delta_f(a) = 1$ if μ_f^a is algebraic and if f is not rational.

The First Main Theorem implies

$$N_f^a(r, s) / T_f(r, s) \leqslant 1 + m_f^a(s) / T_f(r, s)$$

if $0 < s < r < \infty$. Here $m_f^a(s)\,\ddot{\omega}^n(a)$ is integrable $\mathbf{P}(V^*)$. The Lebesgue dominance theorem implies

$$1 \geqslant \int_{\mathbf{P}(V^*)} (1 - \delta_f(a))\,\omega^n(a) \geqslant \limsup_{r\to\infty} \int_{\mathbf{P}(V^*)} \frac{N_f^a(r, s)}{T_f(r, s)}\,\omega^n(a) = 1.$$

Therefore $\delta_f(a) = 0$ for almost all $a \in \mathbf{P}(V^*)$. Consequently, $f(W) \cap \ddot{E}[a] \neq 0$ for almost all $a \in \mathbf{P}(V^*)$. If f is rational, then μ_f^a is algebraic for all $a \in \mathbf{P}(V^*)$. If f is not rational, then μ_f^a is algebraic at most for a set of measure zero in $\mathbf{P}(V^*)$.

Under some additional assumptions, the Second Main Theorem sharpens these results. At first a definition. A subset B of $\mathbf{P}(V^*)$ is said to be basic if and only if there exists a base $\alpha_0, \cdots, \alpha_n$ of V^* such that

$$B = \{\mathbf{P}(\alpha_0), \cdots, \mathbf{P}(\alpha_n)\}.$$

Obviously, $\#B = n + 1$ if B is basic. A subset Q of $\mathbf{P}(V^*)$ is said to be in general position if and only if every subset of Q consisting of $n + 1$ elements is basic.

THEOREM 4.1. SECOND MAIN THEOREM. *Let* W *and* V *be hermitian vector spaces with* $\dim W = m > 0$ *and* $\dim V = n + 1 > 1$. *Let* $f: W \to \mathbf{P}(V)$ *be a meromorphic map. Take* $Q \subseteq \mathbf{P}(V^*)$. *Suppose that the following assumptions are satisfied*:

1. *The set* Q *is in general position.*
2. *The map* f *is not rational.*
3. *If* $a \in \mathbf{P}(V^*)$, *then* $f(W) \not\subseteq \ddot{E}[a]$.
4. *There exists a vector* $\mathfrak{b} \in W\langle 1\rangle$ *such that the following condition is satisfied*: *Let* $\mathfrak{v}: W \to V$ *be a holomorphic map with* $\mathfrak{v}^{-1}(0) = I_f$ *and* $\mathbf{P} \circ \mathfrak{v} = f$ *on* $W - I_f$. *Define*

$$\mathfrak{v}'(\mathfrak{z}) = \frac{d}{dx}\mathfrak{v}(\mathfrak{z} + x\mathfrak{b})\Big|_{x=0} \quad \textit{for all } \mathfrak{z} \in W.$$

Define $\mathfrak{v}^{(0)} = \mathfrak{v}$ *and* $\mathfrak{v}^{(1)} = \mathfrak{v}'$ *and* $\mathfrak{v}^{(p)} = (\mathfrak{v}^{(p-1)})'$ *if* $p > 1$. *Assume that, for all* $p = 0, 1, \cdots, n$ *and all* $\alpha_0, \cdots, \alpha_p$ *in* V^* *with* $\alpha_0 \wedge \cdots \wedge \alpha_p \neq 0$,

$$\det(\alpha_\nu \circ \mathfrak{v}^{(\mu)}) \not\equiv 0.$$

Then the set $\{a \in Q \mid \delta_f(a) > 0\}$ *is at most countable and*

$$\sum_{a\in Q} \delta_f(a) \leqslant n + 1. \tag{4.4}$$

The proof is easily obtained from [49, §24]. Under these assumptions, f intersects at least one among $n + 2$ hyperplanes in general positions and the counter image of at least one is not algebraic.

The Second Main Theorem was obtained by Ahlfors [1] for $m = 1$ and by Weyl [66] for maps on a Riemann surface. For $m = n > 0$ another Second Main Theorem was proved by Carlson and Griffiths [13], and for $m \geq n$ by Griffiths and King [17].

If $f: W \to \mathbf{P}(V)$ is a meromorphic map but not holomorphic, then $f(W) \cap \ddot{E}[a] \neq \emptyset$ for all $a \in \mathbf{P}(V^*)$. Let $Q \neq \emptyset$ be a finite subset of $\mathbf{P}(V^*)$. Define

$$E = \bigcup_{a \in Q} \ddot{E}[a].$$

Green [16] obtains the following results:

1. If Q is in general position with $\#Q = n + k$, if $k \geq 2$ and $q = [n/k]$, then there exists a holomorphic map

$$f: \mathbf{C}^q \longrightarrow \mathbf{P}(V^*) - E$$

such that $f(\mathbf{C}^q)$ is open and dense in a projective plane of dimension q in $\mathbf{P}(V^*)$.

2. If Q is in general position with $\#Q = n + k$, if $k \geq 2$ and if $f: W \to \mathbf{P}(V^*) - E$ is holomorphic, then $f(W)$ is contained in a projective plane of dimension q with $q \leq [n/k]$.

3. If $\#Q = n + 2$, and if $f: W \to \mathbf{P}(V^*) - E$ is holomorphic, then $f(W)$ is contained in a hyperplane of $\mathbf{P}(V^*)$.

Now the case of meromorphic and holomorphic functions shall be considered. Let f be a meromorphic function on $W(R)$ with $0 < R \leq +\infty$. Then $f: W(R) \longrightarrow P_1$ is a meromorphic map. Hence a holomorphic vector function $\mathfrak{v}: W(R) \to \mathbf{C}^2$ exists such that $\mathfrak{v}^{-1}(0) = I_f$ and $\mathbf{P} \circ \mathfrak{v} = f$ on $W(R) - I_f$. Observe $\mathfrak{v} = (h, g)$ where h and g are holomorphic functions which are coprime at every point of $W(R)$ such that $h \cdot f = g$. If $0 < s < r < R$, then

$$T_f(r, s) = \frac{1}{2} \int_{W\langle r \rangle} \log (|g|^2 + |h|^2)\sigma - \frac{1}{2} \int_{W\langle 1 \rangle} \log (|g|^2 + |h|^2)\sigma.$$

Assume that f is holomorphic. Then $h \equiv 1$ and $g = f$ can be taken. Therefore

$$T_f(r) = \int_{W\langle r \rangle} \log \sqrt{1 + |f|^2}\, \sigma - \log \sqrt{1 + |f(0)|^2}.$$

For $0 \leq r \leq R$, define

$$M_f(r) = \sup \{ |f(\mathfrak{z})| \mid |\mathfrak{z}| < r \}.$$

Theorem 3.8 easily implies

THEOREM 4.2. *Take $0 < \theta < 1$ and $0 < r < R \leq +\infty$. Let $f: W(R) \to \mathbf{C}$ be a holomorphic function. Then*

$$T_f(r) \leq \log^+ M_f(r) + \tfrac{1}{2} \log r,$$

$$\log^+ M_f(\theta r) \leq \frac{1 + \theta}{(1 - \theta)^{2m-1}} [T_f(r) + \tfrac{1}{2} \log (1 + |f(0)|^2)].$$

(Compare Kneser [21] and [47, Satz 2].)

5. **The Jensen-Poisson Formula. Canonical Functions**

Again, let W be a hermitian vector space of dimension $m > 1$. Take $0 < R \leqslant +\infty$. Let f be a meromorphic function on $W(R)$. Assume that f is holomorphic at $0 \in W$ with $f(0) \neq 0$. Abbreviate $\nu = \mu_f$ and $A = \operatorname{supp} \nu$. Assume $A \neq \varnothing$. Take s maximal with $A(s) = \varnothing$. Take $\mathfrak{a} \in W\langle 1\rangle$ and $s \leqslant r < R$. For $z \in \mathbf{C}(s)$, the Jensen-Poisson formula states

$$\begin{aligned} \log \frac{f(z\mathfrak{a})}{f(0)} &= \frac{1}{2\pi}\int_0^{2\pi} \log |f(re^{i\phi}\mathfrak{a})| \frac{2z}{re^{i\phi} - z}\, d\phi \\ &+ \sum_{u \in \mathbf{C}(r)} \nu(u;\mathfrak{a}) \log(1 - z/u) \\ &- \sum_{u \in \mathbf{C}(r)} \nu(u;\mathfrak{a}) \log(1 - z\bar{u}/r^2). \end{aligned}$$

Take $\mathfrak{z} \in W(s)$. Then $|(\mathfrak{z}|\mathfrak{a})| \leqslant |\mathfrak{z}| < s$. Substitute $z = (\mathfrak{z}|\mathfrak{a})$ into (5.1). Observe

$$\frac{2z}{re^{i\phi} - z} = \frac{2(\mathfrak{z}|re^{i\phi}\mathfrak{a})}{r^2 - (\mathfrak{z}|re^{i\phi}\mathfrak{a})},$$

$$z/u = (\mathfrak{z}|u\mathfrak{a})/(u\mathfrak{a}|u\mathfrak{a}), \qquad z\bar{u} = (\mathfrak{z}|u\mathfrak{a}).$$

With these substitutions, (5.1), Lemma 2.2 and Theorem 3.6 imply

$$\begin{aligned} &\int_{W\langle 1\rangle} \log \frac{f((\mathfrak{z}|\mathfrak{a})\mathfrak{a})}{f(0)}\, \sigma(\mathfrak{a}) \\ &= \int_{W\langle r\rangle} \log |f(\mathfrak{y})| \frac{2(\mathfrak{z}|\mathfrak{y})}{r^2 - (\mathfrak{z}|\mathfrak{y})}\, \sigma(\mathfrak{y}) \\ &+ \int_{A(r)} \nu(\mathfrak{y}) \log(1 - (\mathfrak{z}|\mathfrak{y})/(\mathfrak{y}|\mathfrak{y}))\, \omega^{m-1} \\ &- \int_{A(r)} \nu(\mathfrak{y}) \log(1 - (\mathfrak{z}|\mathfrak{y})/r^2)\, \omega^{m-1} \end{aligned} \tag{5.2}$$

for all $\mathfrak{z} \in W(s)$. The first integral can be evaluated by the following fundamental lemma on pluri-harmonic and pluri-subharmonic functions.

LEMMA 5.1. *Let* $0 < R \leqslant +\infty$. *Let* u *be a pluri-subharmonic on* $W(R)$. *Let* $g : \mathbf{R}[0, 1] \to \mathbf{R}_+$ *be a bounded, measurable function. Take* $\mathfrak{z} \in W(R)$. *Then*

$$\begin{aligned} &(m-1)\int_0^1 (1-t)^{m-2} g(t) u(t\mathfrak{z})\, dt \\ &\leqslant \int_{W\langle 1\rangle} g(|(\mathfrak{z}|\mathfrak{a})|^2/|\mathfrak{z}|^2) u((\mathfrak{z}|\mathfrak{a})\mathfrak{a})\sigma(\mathfrak{a}). \end{aligned} \tag{5.3}$$

If $u : W(R) \to \mathbf{R}$ is pluri-harmonic, or if $u : W(R) \to \mathbf{C}$ is holomorphic, equality holds in (5.3).

Therefore

(5.4)
$$\int_{W\langle 1\rangle} \log \frac{f((\mathfrak{z}|\mathfrak{a})\mathfrak{a})}{f(0)}\,\sigma(\mathfrak{a}) = (m-1)\int_0^1 (1-t)^{m-2} \log \frac{f(\mathfrak{z}t)}{f(0)}\,dt$$

for all $\mathfrak{z} \in W(s)$. (The proof proceeds as in [47, Hilfssatz 6].)

Let $h: W(R) \to \mathbf{C}$ be a function. Take any $p \in \mathbf{N}$. Define

$$(I_p h)(\mathfrak{z}) = p\int_0^1 (1-t)^{p-1} h(t\mathfrak{z})\,dt$$

for $\mathfrak{z} \in W(R)$ provided the integral exists. If h is holomorphic on $W(R)$, then

$$(I_p h)(x\mathfrak{z}) = \frac{p!}{x^p}\int_0^x \int_0^{x_2} \cdots \int_0^{x_{p-1}} h(x_p\mathfrak{z})\,dx_p \cdots dx_1$$

for all $\mathfrak{z} \in W(R)$ and $x \in \mathbf{C} - \{0\}$ with $|x\mathfrak{z}| < R$. Define

$$(D_p h)(\mathfrak{z}) = \frac{1}{p!}\frac{d^p}{dx^p}\,[x^p h(x\mathfrak{z})]\,\Big|_{x=1}.$$

Then

$$D_p \circ I_p h = h = D_p \circ I_p h.$$

Let $\mathfrak{H}(R)$ be the vector space of all holomorphic functions on $W(R)$. Then

$$I_p = D_p^{-1} : \mathfrak{H}(R) \longrightarrow \mathfrak{H}(R)$$

is a linear isomorphism.

Define the holomorphic function $L: \mathbf{C}(1) \to \mathbf{C}$ by

(5.5)
$$L(z) = \frac{1}{(m-1)!}\frac{d^{m-1}}{dz^{m-1}}\,(z^{m-1}\log(1-z)) = D_{m-1}\log(1-z)$$

for $z \in \mathbf{C}(1)$. Under consideration of (5.4) apply D_{m-1} to (5.1). This yields the following theorem.

THEOREM 5.2. JENSEN-POISSON FORMULA ([45, [53]). *Take $0 < R \leqslant +\infty$. Let f be a meromorphic function on $W(R)$ which is holomorphic at $0 \in W$ with $f(0) \neq 0$. Define $\nu = \mu_f$ and $A = \operatorname{supp} \mu_f$. Assume $A \neq \varnothing$. Let s be maximal with $0 < s < R$ such that $A(s) = \varnothing$. Take $r \in \mathbf{R}[s, R)$. Then holomorphic functions F and E on $W(r)$ are defined by*

(5.6)
$$F(\mathfrak{z}) = \int_{W\langle r\rangle} \log|f(\mathfrak{y})|\left[\frac{r^{2m}}{(r^2 - (\mathfrak{z}|\mathfrak{y}))^m} - 1\right]\sigma(\mathfrak{y}),$$

(5.7)
$$E(\mathfrak{z}) = \int_{A(r)} \nu(\mathfrak{y}) L((\mathfrak{z}|\mathfrak{y})/r^2)\,\omega^{m-1}(\mathfrak{y}),$$

for all $\mathfrak{z} \in W(r)$. Also a holomorphic function H on $W(s)$ is defined by

(5.8)
$$H(\mathfrak{z}) = \int_{A(r)} \nu(\mathfrak{y}) L((\mathfrak{z}|\mathfrak{y})/(\mathfrak{y}|\mathfrak{y}))\,\omega^{m-1}(\mathfrak{y}).$$

Moreover, for $\mathfrak{z} \in W(s)$ the following identity holds:

(5.9)
$$\log f(\mathfrak{z})/f(0) = F(\mathfrak{z}) + H(\mathfrak{z}) - E(\mathfrak{z}).$$

The Jensen-Poisson formula (5.9) leads to the construction of a canonical function. Let ν be a divisor on $W(R)$ such that $0 \notin \operatorname{supp} \nu = A \neq \varnothing$. Let s be the biggest number such that $A(s) = \varnothing$. Take r with $s \leqslant r < R$. Then a holomorphic function $H: W(r) \to \mathbf{C}$ is defined by (5.8). Now, take any meromorphic function f on $W(R)$ with $\mu_f = \nu$. Define F and E as holomorphic functions on $W(r)$ by (5.6) and (5.7) respectively. Then

$$h = (f/f(0))e^{E-F}$$

is a meromorphic function on $W(r)$ with $\mu_h = \mu_f \,|\, W(r)$. Also

$$h = e^H \quad \text{on } W(s).$$

Therefore h does not dependon the choice of f but only on the divisor ν and the number r. Observe $H(0) = 0$ and $h(0) = 1$. The first part of the following theorem is proved.

THEOREM 5.3 [53]. *Take $0 < R \leqslant +\infty$. Let ν be a divisor on $W(R)$. Define $A = \operatorname{supp} \nu$. Assume $0 \notin A \neq \varnothing$. Let s be maximal with $0 < s < R$ such that $A(s) = \varnothing$. Take r with $s \leqslant r < R$. Then one and only one meromorphic function h on $W(r)$ exists such that $h(0) = 1$, such that h is holomorphic on $W(s)$ with $h(\mathfrak{z}) \neq 0$ for all $\mathfrak{z} \in W(s)$ and such that*

$$\log h(\mathfrak{z}) = \int_{A(r)} \nu(\mathfrak{y}) L((\mathfrak{z}\,|\,\mathfrak{y})/(\mathfrak{y}\,|\,\mathfrak{y}))\,\omega^{m-1}(\mathfrak{y})$$

for all $\mathfrak{z} \in W(s)$. Moreover, $\mu_h = \nu \,|\, W(r)$. If $\nu \geqslant 0$, then h is holomorphic on $W(r)$, and if $0 < \theta < 1$, then

$$\log M_h(\theta^3 r) \leqslant \frac{8m}{(1-\theta)^{3m}} \,[N_\nu(r) + n_\nu(r)]\,. \tag{5.10}$$

Only (5.10) remains to be proved, which is based on the following lemma.

LEMMA 5.4. *Assume $0 < \theta < 1$ and $0 < r < R \leqslant +\infty$. Let f be a holomorphic function on $W(R)$ with $f(0) = 1$. Define $\nu = \mu_f$. Take $p \in \mathbf{N}$. Define $Q = I_p(\log|f|)$, i. e.,*

$$Q(\mathfrak{z}) = p \int_0^1 (1-t)^{p-1} \log|f(t\mathfrak{z})|\, dt.$$

Define

$$K_f(r) = \operatorname{Max}\,\{Q(\mathfrak{z}) \,|\, \mathfrak{z} \in W[r]\}.$$

Then

$$\log M_f(\theta^3 r) \leqslant \frac{4}{(1-\theta)^{p+1+2m}} \,[K_f(r) + (2p-1)\,n_\nu(r) + N_\nu(r)]\,.$$

(See [47, Hilfssatz 7].)

PROOF OF (5.10). Take any $\mathfrak{a} \in W\langle 1\rangle$. Then

$$h(z\mathfrak{a}) = e^{Q(z;\mathfrak{a})} \prod_{0 \neq u \in \mathbf{C}(r)} (1 - z/u)^{\nu(u;\mathfrak{a})} \tag{5.11}$$

for all $z \in \mathbf{C}(r)$. Here $Q(z;\mathfrak{a})$ is a holomorphic function of z on $\mathbf{C}(r)$ for each fixed $\mathfrak{a} \in W\langle 1\rangle$ uniquely determined by $Q(0;\mathfrak{a}) = 0$. Take any fixed $\mathfrak{z} \in W(r)$. Then

$|(\mathfrak{z}|\mathfrak{a})| < r$. Then, as shown in Mueller [31], $Q((\mathfrak{z}|\mathfrak{a});\mathfrak{a})$ is a measurable function of $\mathfrak{a}$ on $W\langle 1\rangle$ which is majorized uniformly on each compact subset $W(r)$ by a function integrable over $W\langle 1\rangle$. Therefore a holomorphic function P on $W(r)$ is defined as

$$P(\mathfrak{z}) = \int_{W\langle 1\rangle} Q((\mathfrak{z}|\mathfrak{a});\mathfrak{a})\,\sigma(\mathfrak{a})$$

for all $\mathfrak{z} \in W(r)$.

Now, $P \equiv 0$ shall be shown: Take $\mathfrak{z} \in W(s)$. Then (5.11) implies

$$\begin{aligned}(m-1)&\int_0^1 (1-t)^{m-2} \log h(t\mathfrak{z})\,dt\\ &= \int_{W\langle 1\rangle} \log h((\mathfrak{z}|\mathfrak{a})\mathfrak{a})\,\sigma(\mathfrak{a})\\ &= P(\mathfrak{z}) + \int_{W\langle 1\rangle} \sum_{0\neq u\in \mathbf{C}(R)} \nu(u;\mathfrak{a}) \log\left(1 - \frac{(\mathfrak{z}|u\mathfrak{a})}{(u\mathfrak{a}|u\mathfrak{a})}\right)\sigma(\mathfrak{a})\\ &= P(\mathfrak{z}) + \int_{A(r)} \nu(\mathfrak{y}) \log\left(1 - \frac{(\mathfrak{z}|\mathfrak{y})}{(\mathfrak{y}|\mathfrak{y})}\right)\omega^{m-1}(\mathfrak{y}).\end{aligned}$$

Apply the differential operator D_{m-1}. Then

$$\log h = D_{m-1}P + \log h \quad \text{on} \quad W(s).$$

Therefore $D_{m-1}P \equiv 0$ on $W(s)$. Hence $D_{m-1}P \equiv 0$ on $W(r)$ and $P = I_{m-1}D_{m-1}P \equiv 0$ on $W(r)$.

Take $\mathfrak{z} \in W(r)$. Then (5.11) implies

$$\begin{aligned}\log &|h((\mathfrak{z}|\mathfrak{a})\mathfrak{a})|\\ &\leqslant \Re\, Q((\mathfrak{z}|\mathfrak{a});\mathfrak{a}) + \sum_{u\in\mathbf{C}(r)} \nu(u;\mathfrak{a}) \log|1 - (\mathfrak{z}|\mathfrak{a})/u|\\ &\leqslant \Re\, Q((\mathfrak{z}|\mathfrak{a});\mathfrak{a}) + \sum_{u\in\mathbf{C}(r)} \nu(u;\mathfrak{a}) \log(1 + r/|u|)\\ &\leqslant \Re\, Q((\mathfrak{z}|\mathfrak{a});\mathfrak{a}) + n_\nu(r;\mathfrak{a}) + N_\nu(r;\mathfrak{a}).\end{aligned}$$

Here

$$\int_{W\langle 1\rangle} \Re\, Q((\mathfrak{z}|\mathfrak{a});\mathfrak{a})\,\sigma_1(\mathfrak{a}) = \Re\, P(\mathfrak{z}) = 0.$$

Lemma 5.1 with $g \equiv 1$ shows

$$\begin{aligned}(m-1)&\int_0^1 (1-t)^{m-2} \log|h(t\mathfrak{z})|\,dt\\ &\leqslant \int_{W\langle 1\rangle} \log|h((\mathfrak{z}|\mathfrak{a})\mathfrak{a})|\,\sigma(\mathfrak{a}) \leqslant n_\nu(r) + N_\nu(r).\end{aligned}$$

This estimate and Lemma 5.4 imply (5.10). q. e. d.

In [53], Theorem 5.3 was used to establish a theory of normal families of nonnegative divisors. The function $h = h_\nu$ depends continuously on the divisor in the appropriate topology on the set of all nonnegative divisors. A shortcoming, so far, is the restriction $r < R$. With weights $r = R$ can be achieved. Only two cases are of interest $R = \infty$ and $R = 1$. Here the case $R = \infty$ shall be considered first. The case $R = 1$ shall be deferred to §9.

Define $\exp(z) = e^z$ for all $z \in \mathbf{C}$. For $q \in \mathbf{Z}_+$ and $z \in \mathbf{C}$, define the Weierstrass prime factor by

$$E(z, q) = (1 - z)\exp\left(\sum_{\lambda=1}^{q} \frac{z^\lambda}{\lambda}\right).$$

For $m \in \mathbf{N}$ define $e_m = D_{m-1} \log E$, that is

$$e_m(z, q) = \frac{1}{(m-1)!} \frac{d^{m-1}}{dz^{m-1}} (z^{m-1} \log E(z, q)),$$

$$e_m(z, q) = -\sum_{p=q+1}^{\infty} \binom{p+m-1}{p} \frac{z^p}{p},$$

$$|e_m(z, q)| \leqslant |z|^{q+1}(1 - |z|)^{-m},$$

for all $z \in \mathbf{C}(1)$.

Let $\nu \geqslant 0$ be a nonnegative divisor on the hermitian vector space W. Define $A = \operatorname{supp} \nu$. Assume $0 \notin A$. A nonnegative, increasing, integral valued function $q: \mathbf{R}_+ \to \mathbf{Z}_+$ is said to be a weight for ν if and only if the integral

$$\int_A \nu(r/\tau_0)^{q\circ\tau_0+1}\, \omega^{m-1} < \infty$$

converges for all $r > 0$. For every given nonnegative divisor ν there exists a weight function for ν [47].

THEOREM 5.5. WEIERSTRASS INTEGRAL THEOREM [47]. *Let $\nu \geqslant 0$ be a nonnegative divisor on W. Define $A = \operatorname{supp} \nu$. Assume $0 \notin A \neq \varnothing$. Define $s = \sup\{r \mid A(r) = \varnothing\}$. Let q be a weight for ν. Then there exists one and only one holomorphic function $h: W \to \mathbf{C}$ with $h(0) = 1$ such that $h(\mathfrak{z}) \neq 0$ for all $\mathfrak{z} \in W(s)$ and such that*

$$\log h(\mathfrak{z}) = \int_A \nu(\mathfrak{y})\, e_m\left(\frac{(\mathfrak{z}\mid\mathfrak{y})}{(\mathfrak{y}\mid\mathfrak{y})}, q(|\mathfrak{y}|)\right) \omega^{m-1}(\mathfrak{y}) \tag{5.12}$$

for all $\mathfrak{z} \in W(s)$. Moreover, $\mu_h = \nu$ is the divisor of h on W.

PROOF. Take any number $r > s$. For $t \geqslant 0$ and $p \in \mathbf{N}$ define

$$\chi_p(t, r) = \begin{cases} 1 & \text{if } p \leqslant q(t), \\ 0 & \text{if } q(t) < p. \end{cases}$$

If $|z| < 1$ and $0 \leqslant t \leqslant r$, then

$$e_m(z, q(t)) - L(z) = \sum_{p=1}^{q(r)} \chi_p(t, r) \binom{p+m-1}{p} \frac{z^p}{p}.$$

Define the polynomial G_r by

$$G_r(\mathfrak{z}) = \sum_{p=1}^{q(r)} \frac{1}{p} \binom{p+m-1}{p} \int_{A(r)} \nu(\mathfrak{y})\, \chi_p(|\mathfrak{y}|, r) \left(\frac{(\mathfrak{z}\mid\mathfrak{y})}{(\mathfrak{y}\mid\mathfrak{y})}\right)^p \omega^{m-1}(\mathfrak{y})$$

for all $\mathfrak{z} \in W$. If $\mathfrak{z} \in W(s)$ and $\mathfrak{y} \in A(r)$, then

$$|(\mathfrak{z}\mid\mathfrak{y})/(\mathfrak{y}\mid\mathfrak{y})| \leqslant |\mathfrak{z}|/|\mathfrak{y}| < 1.$$

Therefore

$$G_r(\mathfrak{z}) = \int_{A(r)} \nu(\mathfrak{y}) e_m \left(\frac{(\mathfrak{z} \mid \mathfrak{y})}{(\mathfrak{y} \mid \mathfrak{y})}, q(|\mathfrak{y}|) \right) \omega^{m-1}(\mathfrak{y})$$
$$- \int_{A(r)} \nu(\mathfrak{y}) L \left(\frac{(\mathfrak{z} \mid \mathfrak{y})}{(\mathfrak{y} \mid \mathfrak{y})} \right) \omega^{m-1}(\mathfrak{y}).$$

A holomorphic function H_r exists on $W(r)$ with $H_r(0) = 1$ such that $\mu_{H_r} = \nu \mid W(r)$ and such that

$$\log H_r(\mathfrak{z}) = \int_{A(r)} \nu(\mathfrak{y}) L \left(\frac{(\mathfrak{z} \mid \mathfrak{y})}{(\mathfrak{y} \mid \mathfrak{y})} \right) \omega^{m-1}(\mathfrak{y})$$

for all $\mathfrak{z} \in W(s)$.

Take any $0 < \theta < 1$. If $\mathfrak{y} \in A - A(r)$ and $\mathfrak{z} \in W[\theta r]$, then $|\mathfrak{z}|\,|\mathfrak{y}|^{-1} \leqslant \theta$. Hence

$$\left| e_m \left(\frac{(\mathfrak{z} \mid \mathfrak{y})}{(\mathfrak{y} \mid \mathfrak{y})}, q(|\mathfrak{y}|) \right) \right| \leqslant \left(\frac{\theta r}{|\mathfrak{y}|} \right)^{q(|\mathfrak{y}|)+1} \frac{1}{(1-\theta)^m}.$$

Therefore

$$B_r(\mathfrak{z}) = \int_{A - A(r)} \nu(\mathfrak{y}) e_m \left(\frac{(\mathfrak{z} \mid \mathfrak{y})}{(\mathfrak{y} \mid \mathfrak{y})}, q(|\mathfrak{y}|) \right) \omega^{m-1}(\mathfrak{y})$$

is defined for each $\mathfrak{z} \in W(r)$. The function B_r is holomorphic on $W(r)$. Observe $B_r(0) = G_r(0) = 0$. On $W(r)$, define

$$h_r = H_r \exp(G_r + B_r).$$

Hence $h_r(0) = 1$ and $\mu_{h_r} = \nu \mid W(r)$. If $\mathfrak{z} \in W(s)$, then

$$\log h_r(\mathfrak{z}) = \int_A \nu(\mathfrak{y}) e_m \left(\frac{(\mathfrak{z} \mid \mathfrak{y})}{(\mathfrak{y} \mid \mathfrak{y})}, q(|\mathfrak{y}|) \right) \omega^{m-1}(\mathfrak{y}).$$

Therefore h_r does not depend on r. If $x > r$, then $h_x \mid W(r) = h_r$. One and only one entire function h exists such that $h \mid W(r) = h_r$. Obviously, $h(0) = 1$ and $\mu_h = \nu$ on W. Also (5.12) holds for all $\mathfrak{z} \in W(s)$. q.e.d.

The function h is said to be the canonical function to ν for the weight q.

6. **Function of Finite Order**

Consider an increasing, nonnegative function $s : \mathbf{R}_+ \to \mathbf{R}_+$. The order of s is defined by

$$\text{Ord } s = \limsup_{r \to \infty} ((\log^+ s(r))/\log r).$$

For $0 < \mu \in \mathbf{R}$,

$$t_\mu(s) = \limsup_{r \to \infty} \frac{s(r)}{r^\mu}, \qquad J_\mu(s) = \int_1^\infty s(t) \frac{dt}{t^{\mu+1}} .$$

Then

$$\begin{aligned} &t_\mu(s) = J_\mu(s) = \infty && \text{if } 0 < \mu < \text{Ord } s, \\ &t_\mu(s) < \infty \text{ and } J_\mu(s) < \infty && \text{if } \mu > \text{Ord } s. \end{aligned}$$

If $0 < \text{Ord } s = \lambda < \infty$, the function s is said to be of

$$\begin{aligned} &\text{maximal type} && \text{iff} \quad t_\lambda(s) = +\infty, \\ &\text{middle type} && \text{iff} \quad 0 < t_\lambda(s) < +\infty, \\ &\text{minimal type} && \text{iff} \quad t_\lambda(s) = 0. \end{aligned}$$

The function s is said to be in the

$$\begin{aligned} &\text{convergence class} && \text{iff} \quad J_\lambda(s) < +\infty, \\ &\text{divergence class} && \text{iff} \quad J_\lambda(s) = +\infty. \end{aligned}$$

If s is of minimal type, then s is in the convergence class.

Suppose that $a > 0$ and that $s : \mathbf{R}[a, +\infty) \to \mathbf{R}_+$ is an increasing, nonnegative function. Define $\tilde{s}(s) = s(x)$ for all $x \geqslant a$ and $\tilde{s}(x) = 0$ for all $x \in \mathbf{R}[0, a)$. Define the order, class, type of s as the order, class and type of $\tilde{s}$. Similarly define $t_\mu(s) = t_\mu(\tilde{s})$ and $J_\mu(s) = J_\mu(\tilde{s})$.

Let $\nu \geqslant 0$ be a nonnegative divisor on W, then the order, class and type of ν is defined as the Ord class type of n_ν. Observe that for each $s > 0$ the order, class and type of n_ν coincide with the order, class and type of $N_\nu(\square, s)$. If $0 \notin \text{supp } \nu$, this is true for $s = 0$ also.

Define $A = \text{supp } \nu$ and $B = A - A(1)$. A number $\mu > 0$ is an exponent of convergence of ν if one – hence all – of the following integrals exist:

$$\int_B \nu \tau_0^{-\mu} \omega^{m-1} < \infty,$$

$$\int_B \nu\tau_0^{1-m-\mu}\upsilon^{m-1} < \infty,$$

$$\int_1^\infty n_\nu(t)t^{-\mu-1}dt < \infty,$$

$$\int_1^\infty N_\nu(t, s)t^{-\mu-1}\,dt < \infty, \quad \text{if } s > 0.$$

Otherwise μ is said to be an exponent of divergence of ν. If $\mu > \operatorname{Ord}\nu$, then μ is an exponent of convergence of ν; if $0 < \mu < \operatorname{Ord}\nu$, then μ is an exponent of divergence of ν. If $0 < \operatorname{Ord}\nu = \mu < \infty$, then μ is an exponent of convergence of ν if and only if ν is in the convergence class.

If $0 \notin A$ and $\mathfrak{a} \in W\langle 1\rangle$, Proposition 3.9 implies $\operatorname{Ord}\nu \geqslant \operatorname{Ord}\nu[\mathfrak{a}]$; and if $\operatorname{Ord}\nu < \infty$, it implies $\operatorname{class}\nu \geqslant \operatorname{class}\nu[\mathfrak{a}]$ and $\operatorname{type}\nu \geqslant \operatorname{type}\nu[\mathfrak{a}]$. Here the following ordering is used:

maximal type > middle type > minimal type,

divergence class > convergence class.

Let W and V be hermitian vector spaces with $\dim W = m$ and $\dim V = n + 1 > 1$. Let $f: W \to \mathbf{P}(V)$ be a meromorphic map. The order, class and type of f are defined as the order, class and type of the characteristic $T_f(r, s)$ which are independent of $s > 0$. If f is holomorphic at $0 \in W$, also $s = 0$ may be included. Take $a \in \mathbf{P}(V^*)$. Assume $f(W) \not\subseteq \ddot{E}[a]$. The First Main Theorem implies $\operatorname{Ord}\mu_f^a \leqslant \operatorname{Ord} f$. If $\operatorname{Ord} f < \infty$, then $\operatorname{class}\mu_f^a \leqslant \operatorname{class} f$ and $\operatorname{type}\mu_f^a \leqslant \operatorname{type} f$. Assume that f is holomorphic at $0 \in W$. Take $\mathfrak{a} \in W\langle 1\rangle$. Define $j_\mathfrak{a}: \mathbf{C} \to W$ by $j_\mathfrak{a}(z) = z\mathfrak{a}$. Then (4.2) implies $\operatorname{Ord} f \circ j_\mathfrak{a} \leqslant \operatorname{Ord} f$. If $\operatorname{Ord} f < \infty$, then $\operatorname{class} f \circ j_\mathfrak{a} \leqslant \operatorname{class} f$ and $\operatorname{type} f \circ j_\mathfrak{a} \leqslant \operatorname{type} f$.

Let $f: W \to \mathbf{C}$ be a holomorphic function. Theorem 4.2 implies $\operatorname{Ord} f = \operatorname{Ord}\log^+ M_f$. If $\operatorname{Ord} f < \infty$, then class $f = \operatorname{class}\log^+ M_f$ and type $f = \operatorname{type}\log^+ M_f$. The following two propositions are easily proved.

PROPOSITION 6.1. *Let $h \not\equiv 0$ be a holomorphic function on W. Assume that* $\operatorname{Ord}(\exp h) = \lambda < +\infty$. *Then h is a polynomial of degree λ. Hence λ is an integer. Also* $\exp h$ *has middle type.*

PROPOSITION 6.2. *Let $g \not\equiv 0$ and $h \not\equiv 0$ be meromorphic functions on W with the same divisor $\mu_h = \mu_g$. Assume $\rho \in \mathbf{R}_+$ exists such that* $\operatorname{Ord} g \leqslant \rho$ *and* $\operatorname{Ord} h \leqslant \rho$. *Then there exists a polynomial Q such that $g = h\exp Q$. Moreover $Q \equiv 0$ or* $\deg Q \leqslant \rho$.

The order, class and type of a nonnegative divisor and of a meromorphic map $f: W \to \mathbf{P}(V)$ are invariant under affine transformations on W and – in case of a meromorphic map – under the automorphisms of $\mathbf{P}(V)$.

Let $s: \mathbf{R}_+ \to \mathbf{R}_+$ be an increasing function. Take $q \in \mathbf{R}_+$. Assume that $J_{q+2}(s) < \infty$. Assume that there exists a number $r_0 > 0$ such that $s(x) = 0$ for all $x \in \mathbf{R}[0, r_0)$. Define

$$K_q(r; s) = r^q \int_0^r s(t)\,t^{-q-1}\,dt + r^{q+1}\int_r^\infty s(t)\,t^{-q-2}\,dt, \tag{6.1}$$

(6.2) $$c(q) = 9(q+1)^2\,(2+\log(1+q)).$$

THEOREM 6.3. CANONICAL FUNCTION OF FINITE ORDER [47]. *Let* ν *be a nonnegative divisor on* W *with* $0 \notin \operatorname{supp} \nu$. *Assume that* $q+1 \in \mathbf{N}$ *is an exponent of convergence of* ν. *Let* h *be the canonical function of* ν *for the constant weight* q *of* ν. *Take* $\theta \in \mathbf{R}$ *and* $r \in \mathbf{R}$ *with* $0 < \theta < 1$ *and* $r > 0$. *Then*

$$\log M_h(\theta r) \leqslant 8m\,c(q)\,(1-\theta)^{-3m}\,K_q(r; n_\nu),$$

$$\operatorname{Ord} \nu \leqslant \operatorname{Ord} h \leqslant \operatorname{Max}(q, \operatorname{Ord} \nu) \leqslant q+1.$$

If $q \in \mathbf{Z}_+$ *is the smallest, nonnegative integer, such that* $q+1$ *is an exponent of convergence of* ν, *then*

$$q \leqslant \operatorname{Ord} \nu = \operatorname{Ord} h \leqslant q+1,$$

$$\operatorname{type} h = \operatorname{type} \nu \quad \text{provided} \quad q < \operatorname{Ord} h \leqslant q+1,$$

$$\operatorname{class} h = \operatorname{class} \nu \quad \text{provided} \quad q < \operatorname{Ord} h < q+1.$$

The theorem was proved in [47]. The proof runs along the lines of the proof of (5.10). To simplify the language, the following definition shall be made. Let $\nu \geqslant 0$ be a nonnegative divisor of finite order on W. Then the smallest nonnegative integer q such that $q+1$ is an exponent of convergence of ν is called the genus of ν.

The canonical function of a nonnegative divisor ν on W for a constant weight q can be intrinsically characterized.

PROPOSITION 6.4 (RONKIN [35, THEOREM 3]). *Let* $\nu \geqslant 0$ *be a nonnegative divisor on* W *with* $0 \notin \operatorname{supp} \nu$. *Let* $q+1 \in \mathbf{N}$ *be an exponent of convergence of* ν. *Then there exists one and only one holomorphic function* $h: W \to \mathbf{C}$ *such that*

$$h(0) = 1, \qquad \mu_{\log h}(0) \geqslant q+1,$$

$$\mu_h = \nu, \qquad \operatorname{Ord} h \leqslant q+1.$$

and such that if $\operatorname{Ord} h = q+1$, *then* h *is of minimal type. This unique function is the canonical function of* ν *for the constant weight* q.

PROOF. The canonical function h of ν for the weight ν trivially satisfies all these conditions. Let f be any other holomorphic functions on W satisfying all these conditions. According to Proposition 6.2, $f = h \exp g$ where g is a polynomial of degree $g \leqslant q+1$ or $g \equiv 0$. Take g such that $g(0) = 0$. Then g is unique. Assume that $g \not\equiv 0$. Define $n = \deg g$. Then

$$n \geqslant \mu_g(0) \geqslant \operatorname{Max}(\mu_{\log f}(0), \mu_{\log h}(0)) \geqslant q+1,$$

$$n \leqslant \operatorname{Max}(\operatorname{Ord} f, \operatorname{Ord} h) = q+1.$$

Therefore $n = q+1$. A constant $c \geqslant 0$ exists such that

$$T_{\exp g} \leqslant T_f + T_g + c.$$

Hence

$$t_{q+1}(T_{\exp g}) \leqslant t_{q+1}(T_f) + t_{q+1}(T_g).$$

If $\operatorname{Ord} f < q+1$, then $t_{q+1}(T_f) = 0$. If $\operatorname{Ord} f = q+1$, then f has minimal type, hence $t_{q+1}(T_f) = 0$. Similarly, $t_{q+1}(T_h) = 0$. Therefore $t_{q+1}(T_{\exp g}) = 0$. Hence $\exp g$ has minimal type contrary to Proposition 6.1. The assumption $g \not\equiv 0$ is wrong. Therefore $g \equiv 0$ and $f = h$. q. e. d.

Although $\dim W = m > 1$ was assumed, the same result is true for $m = 1$, where the canonical function is given by the Weierstrass product.

Let $\nu \geqslant 0$ be a nonnegative divisor on W with $0 \notin \operatorname{supp} \nu$. Assume that $q+1 \in \mathbf{N}$ is an exponent of convergence of ν. Take any $\mathfrak{a} \in W\langle 1\rangle$. According to Proposition 3.9, $q+1$ is an exponent of convergence of the divisor $\nu[\mathfrak{a}]$. The Weierstrass product of $\nu[\mathfrak{a}]$ for the weight q is the canonical function of ν for q. Let $j_{\mathfrak{a}} : \mathbf{C} \to W$ defined by $j_{\mathfrak{a}}(z) = z\mathfrak{a}$. Let h be the canonical function of ν for the weight q. Then $h \circ j_{\mathfrak{a}}$ satisfies the conditions of Proposition 6.4 for $\nu[\mathfrak{a}]$. Therefore the following result is obtained.

PROPOSITION 6.5 [47]. *Let $\nu \geqslant 0$ be a nonnegative divisor on W with $0 \notin \operatorname{supp} \nu$. Assume that $q+1 \in \mathbf{N}$ is an exponent of convergence of ν. Let h be the canonical function of ν for the weight q. Take $\mathfrak{a} \in W\langle 1\rangle$. Take $z \in \mathbf{C}$. Then*

$$h(z\mathfrak{a}) = \prod_{0 \neq y} E(z/y, q)^{\nu(y;\mathfrak{a})}.$$

Lelong [28] constructed a canonical function to a given divisor $\nu \geqslant 0$ on W, which by virtue of Proposition 6.5 coincides with the canonical function constructed in Proposition 6.4. For $u \in \mathbf{C}(1)$ and $v \in \mathbf{R}[-1, +1]$ define

$$\frac{1}{(1 - 2uv + u^2)^{m-1}} = \sum_{p=0}^{\infty} B_p^m(v) u^p$$

where B_p^m is a polynomial in v. If $0 \neq \mathfrak{z} \in W$ and $\mathfrak{y} \in W$ with $|\mathfrak{z}| < |\mathfrak{y}|$ define

$$v = \frac{(\mathfrak{z}\,|\,\mathfrak{y}) + (\mathfrak{y}\,|\,\mathfrak{z})}{2|\mathfrak{y}|\,|\mathfrak{z}|}, \qquad u = \frac{|\mathfrak{z}|^2}{|\mathfrak{y}|^2},$$

then $-1 \leqslant v \leqslant 1$ and $0 < u < 1$ and

$$\left(\frac{|\mathfrak{y}|}{|\mathfrak{z} - \mathfrak{y}|}\right)^{2m-2} = \sum_{p=0}^{\infty} B_p^m(v) u^p.$$

Define the Lelong kernel

$$\lambda_m(\mathfrak{z}, \mathfrak{y}, q) = \frac{1}{2m-2} |\mathfrak{y}|^{2-2m}\left[-|\mathfrak{z} - \mathfrak{y}|^{2-2m} + \sum_{p=0}^{q} B_p^m(v) u^p\right]$$

for $0 \neq \mathfrak{z} \in W$ and $0 \neq \mathfrak{y} \in W$ with $|\mathfrak{z}| < |\mathfrak{y}|$. Define $\lambda_m(0, \mathfrak{y}, q) = 0$.

THEOREM 6.6 (LELONG [25], [28]). *Let $\nu \geqslant 0$ be a nonnegative divisor on W. Define $A = \operatorname{supp} \nu$. Assume $0 \notin A \neq \varnothing$. Let $q+1 \in \mathbf{N}$ be an exponent of convergence of ν. Let h be the canonical function of ν for the weight q. Take $\mathfrak{z} \in W - A$. Then*

$$\log |h(\mathfrak{z})| = \int_A \nu \lambda_m(\mathfrak{z}, \mathfrak{y}, q) v^{m-1}.$$

It should be emphasized again that the function h is constructed independently by Lelong and that only Proposition 6.4 gives the identification with the canonical function of Theorem 6.3. It would be of interest to find a way which transforms the integral in Theorem 6.3 to the integral in Theorem 6.6.

The existence of the canonical function can be used for several applications. For instance for algebraic divisors.

PROPOSITION 6.7 ([50], RUTISHAUSER [39]). *A divisor $\nu \geqslant 0$ on W is algebraic if and only if n_ν is bounded. Moreover, if $n_\nu(r) \to n$ for $r \to \infty$ with $n < \infty$, then ν is the divisor of a polynomial of degree n.*

PROOF. If h is a polynomial of degree n such that $\nu = \mu_h$, then $h = h_0 + \cdots + h_n$ where h_p is a homogeneous polynomial of degree p. Here $h_n \not\equiv 0$. If $\mathfrak{a} \in W\langle 1\rangle$ and $h_n(\mathfrak{a}) \neq 0$, then

$$n_\nu(r; \mathfrak{a}) \longrightarrow n \quad \text{for } r \longrightarrow \infty$$

since $h_n(\mathfrak{a}) \neq 0$ for almost all $\mathfrak{a} \in W\langle 1\rangle$. Hence

$$n_\nu(r) = \int_{W\langle 1\rangle} n_\nu(r; \mathfrak{a})\,\sigma(\mathfrak{a}) \longrightarrow n$$

for $r \to \infty$. Especially, n_ν is bounded.

Assume that ν is given and that a constant $C > 0$ exists such that $n_\nu(r) \leqslant C$ for all $r > 0$. If $\mu > 0$, then

$$\int_1^\infty n_\nu(t) \frac{dt}{t^{\mu+1}} < \infty.$$

W.l.o.g. $0 \notin \operatorname{supp} \nu$ can be assumed. The canonical function h of ν exists for the weight q. If $r > 1$, then

$$K_0(r, n_\nu) \leqslant N_\nu(1) + C \log r + C.$$

According to Theorem 6.3, a constant $C_1 > 0$ exists such that for $0 < \theta < 1$ and all $r \geqslant 2$ the following estimate holds:

$$\log M_h(\theta^3 r) \leqslant C_1 \log r$$

which implies that h is a polynomial. q.e.d.

Let V be a hermitian vector space of dimension $n + 1$. Let $f: W \to \mathbf{P}(V)$ be a meromorphic map. Assume that $f(W) \not\subseteq \ddot{E}[a]$ for all $a \in \mathbf{P}(V^*)$. Let $\mathfrak{v}: W \to V$ be a holomorphic map such that $\dim \mathfrak{v}^{-1}(0) \leqslant m - 2$ and such that $f(\mathfrak{z}) = \mathbf{P}(\mathfrak{v}(\mathfrak{z}))$ if $\mathfrak{v}(\mathfrak{z}) \neq 0$. Let $B = \{a_0, \cdots, a_n\}$ be a basic set in $\mathbf{P}(V^*)$. Take $\alpha_\lambda \in \mathbf{P}^{-1}(a_\lambda)$ for $\lambda = 0, \cdots, n$. Then $\alpha_0, \cdots, \alpha_n$ is a base of V^*. Define

$$\nu_\lambda = \mu_f^{a_\lambda} = \mu_{\alpha_\lambda} \circ \mathfrak{v}.$$

Define $A_\lambda = \operatorname{supp} \nu_\lambda$. Assume $0 \notin A_\lambda$. Assume $\operatorname{Ord} f = \rho < \infty$. Let q_λ be the genus of ν_λ. Let h_λ be the canonical function of ν_λ for the weight q_λ. Then $\alpha_\lambda \circ \mathfrak{v} = \exp(g_\lambda) h_\lambda$ where g_λ is an entire function. It can be shown that each function

$$\frac{\alpha_\lambda \circ \mathfrak{v}}{\alpha_\mu \circ \mathfrak{v}} = \exp(g_\lambda - g_\mu) \frac{h_\lambda}{h_\mu} \tag{6.1}$$

has finite order. Hence $g_\lambda - g_\mu$ is a polynomial of degree $p_{\lambda\mu}$. (If $g_\lambda = g_\mu$, set $p_{\lambda\mu} = 0$.) Define

$$q_B = \operatorname{Max} \{p_{\lambda\mu}, q_\sigma \mid \lambda, \mu, \sigma = 0, \cdots, n\}$$

as the genus of f for the basic set B. The number q_B is independent of the special choice of $\mathfrak{v}, \alpha_0, \cdots, \alpha_n$ and depends on f and B only. Also q_B is invariant under affine transformations of W. So q_B can be defined for each basic subset B of $\mathbf{P}(V^*)$.

THEOREM 6.8 [47]. *Let* W *and* V *be hermitian vector spaces of dimension* m *and* $n+1$ *respectively. Let* $f: W \longrightarrow \mathbf{P}(V)$ *be a meromorphic map of positive finite order* $\lambda = \operatorname{Ord} f$. *Then*

1. *For each basic set* $B \subseteq \mathbf{P}(V^*)$, *the following inequality holds*: $q_B \leqslant \lambda \leqslant q_B + 1$.

2. *If* λ *is not an integer, then* q_B *does not depend on the basic set* $B \subseteq \mathbf{P}(V^*)$. *There exists a linear subspace* Y *of* V^* *with* $\dim V^* \leqslant n$ *such that* $\operatorname{Ord} \mu_f^a = \lambda$ *for all* $a \in \mathbf{P}(V^*) - \mathbf{P}(Y)$.

3. *Assume* λ *is an integer. Assume* B *is a basic set. Then* $\lambda = q_B$ *if and only if either* (a) *or* (b) *hold*:

(a) f *does not have minimal type.*

(b) f *has minimal type, belongs to the divergence class and* λ *is an exponent of divergence for at least one* $a \in B$.

4. *Assume* λ *is an integer. Assume* B *is a basic set. Then* $\lambda = q_B + 1$ *if and only if* (c) *or* (d) *hold.*

(c) f *has minimal type, belongs to the divergence class and* λ *is an exponent of convergence for all* $a \in B$.

(d) f *is in the convergence class.*

7. Theta Functions

An additive subgroup $\mathfrak{L}$ of the hermitian vector space W is said to be a lattice in W if and only if $\mathfrak{L}$ generates W over $\mathbf{R}$. For each $\mathfrak{p} \in W$ define $t_{\mathfrak{p}}: W \to W$ by $t_{\mathfrak{p}}(\mathfrak{z}) = \mathfrak{z} + \mathfrak{p}$ for all $\mathfrak{z} \in W$. Let f be a function on W. Then $\mathfrak{p} \in W$ is said to be a period of f if and only if $f \circ t_{\mathfrak{p}} = f$. The set $\mathfrak{P}(f)$ of all periods of f is a submodule of W. The function f is said to be abelian, if and only if $\mathfrak{P}(f)$ is a lattice in W. So, abelian meromorphic functions are defined. An abelian holomorphic function is constant. A divisor $\nu: W \to \mathbf{Z}$ is a function. Hence, an abelian divisor is defined. If $f \not\equiv a$ is an abelian meromorphic function, then μ_f^a is an abelian divisor. If $f \not\equiv 0$, then μ_f is an abelian divisor.

LEMMA 7.1 [48]. *Let $\nu \geqslant 0$ be a nonnegative abelian divisor with $\nu \not\equiv 0$. Then* Ord $\nu = 2$ *and ν belongs to the divergence class ν has middle type.*

The proof is obtained by parallel translation of the fundamental period parallelogram.

Let $f: W \to \mathbf{C}$ be a holomorphic function. Then $\mathfrak{p} \in W$ is said to be a theta period of f if and only if a polynomial $L_{\mathfrak{p}}$ with $\deg L_{\mathfrak{p}} \leqslant 1$ exists such that $f \circ t_{\mathfrak{p}} = f \exp(L_{\mathfrak{p}})$ on W. The set $\mathfrak{Q}(f)$ of all theta periods of f is a submodule of W. Obviously, $\mathfrak{Q}(f) \subseteq \mathfrak{P}(\mu_f)$. A holomorphic function $f \not\equiv 0$ on W is said to be a theta function if and only if $\mathfrak{Q}(f)$ is a lattice in W. The divisor of a theta function is abelian. A theta function f on W is trivial if and only if f^{-1} is also a theta function on W, that is, if and only if $f(\mathfrak{z}) \neq 0$ for all $\mathfrak{z} \in W$.

LEMMA 7.2 [48]. *Let f be a theta function on W. Then* $\operatorname{Ord} f \leqslant 2$. *If f is not trivial, then* $\operatorname{Ord} f = 2$. *Moreover f is in the divergence class and f has middle type.*

The proof is obtained by parallel translation of the fundamental period parallelogram.

LEMMA 7.3 [48]. *A holomorphic function f on W is a trivial theta function, if and only if $f = e^Q$ where $Q \equiv 0$ or where Q is a polynomial with* $\deg Q \leqslant 2$.

PROOF. Trivially, e^Q is a trivial theta function if Q is a polynomial with $Q \equiv 0$ or $\deg Q \leqslant 2$. Let f be a trivial theta function. Then $\operatorname{Ord} f \leqslant 2$ by Lemma 7.2. According to Proposition 6.2, $f = e^Q$, where Q is a polynomial with $Q \equiv 0$ or $\deg Q \leqslant 2$. q.e.d.

A well-known result states that every abelian divisor $\nu \geqslant 0$ is the divisor of a theta function h such that $\mathfrak{Q}(h) = \mathfrak{P}(\nu)$ and $\mu_h = \nu$. This can be proved using the following two lemmata.

LEMMA 7.4 [48]. *Let g, f_1 and f_2 be holomorphic functions on $W(R)$ where*

$0 < R \leqslant +\infty$. *Assume that* $f_1(0) = f_2(0) = 1$ *and* $g(0) = 0$. *Assume that* $f_1 = f_2 \exp(g)$. *Take real numbers* θ *and* r *with* $0 < \theta < 1$ *and* $0 < r < R$. *Then*

$$M_g(\theta r) \leqslant \frac{48}{(1-\theta)^2} (\log M_{f_1}(r) + \log M_{f_2}(r)).$$

LEMMA 7.5 [48]. *Let* $f: W \to \mathbf{C}$ *be a holomorphic function with* $\operatorname{Ord} f \leqslant 2$. *Assume that* $f \not\equiv 0$ *and* $\nu = \mu_f \not\equiv 0$. *Then* $\mathfrak{P}(\nu) = \mathfrak{Q}(f)$.

PROOF. Obviously, $\mathfrak{P}(\nu) \supseteq \mathfrak{Q}(f)$. W.l.o.g. $f(0) = 1$ can be assumed. Take $0 \neq \mathfrak{p} \in \mathfrak{P}(\nu)$. Then $f \circ t_{\mathfrak{p}}$ and f have the same zero divisor. Hence a holomorphic function g on W exists such that $f \circ t_{\mathfrak{p}} = f \exp(g)$. Because $\operatorname{Ord} f \circ t_{\mathfrak{p}} = \operatorname{Ord} f \leqslant 2$, the function g is a polynomial with $g \equiv 0$ or $\deg g \leqslant 2$. Hence $g = c + L + Q$ where c is a constant, where $L: W \to \mathbf{C}$ is linear and where Q is a homogeneous polynomial of degree 2. Now, $Q \equiv 0$ has to be shown.

Take any $n \in \mathbf{N}$. By induction

$$(7.1) \qquad f(\mathfrak{z} + n\mathfrak{p}) = f(\mathfrak{z}) \exp\left(\sum_{\lambda=0}^{n-1} g(\mathfrak{z} + \lambda\mathfrak{p})\right)$$

for all $\mathfrak{z} \in W$. Since $f(0) = 1$, also $f(n\mathfrak{p}) \neq 0$. Define h_n and A_n on W by

$$(7.2) \qquad f_n(\mathfrak{z}) = f(\mathfrak{z} + n\mathfrak{p})/f(n\mathfrak{p}),$$

$$(7.3) \qquad A_n(\mathfrak{z}) = \sum_{\lambda=0}^{n-1} [g(\mathfrak{z} + \lambda\mathfrak{p}) - g(\lambda\mathfrak{p})].$$

Then

$$(7.4) \qquad f_n = f \exp(A_n)$$

with $f_n(0) = 1 = f(0)$ and $A_n(0) = 0$. Take $r > 0$. Take $\mathfrak{z} \in W[r/2]$. Lemma 7.4 implies

$$(7.5) \qquad |A_n(\mathfrak{z})| \leqslant 192 [\log M_{f_n}(r) + \log M_f(r)].$$

By assumption

$$\limsup_{r \to \infty} (\log^+ \log M_f(r))/(\log r) \leqslant 2.$$

Take any fixed number μ with $2 < \mu < 3$. A constant $C_1 > 0$ exists such that

$$(7.6) \qquad \log M_f(r) \leqslant C_1(1 + r)^{\mu}.$$

Take any $\mathfrak{z} \in W[r]$. Then (7.6) implies

$$(7.7) \qquad \log |f(\mathfrak{z} + n\mathfrak{p})| \leqslant C_1(1 + r + n|\mathfrak{p}|)^{\mu}$$

for all $n \in \mathbf{N}$. Constants $C_2 > 0$ and $C_3 > 0$ exist such that

$$(7.8) \qquad |g(\mathfrak{z})| \leqslant C_2 |\mathfrak{z}|^2 + C_3$$

for all $\mathfrak{z} \in W$. Therefore, (7.1) for $\mathfrak{z} = 0$ and (7.8) imply

$$(7.9)\qquad \begin{aligned} |\log|f(n\mathfrak{p})|| &= \left|\Re \sum_{\lambda=0}^{n-1} g(\lambda\mathfrak{p})\right| \\ &\leqslant \sum_{\lambda=0}^{n-1} C_2\lambda^2|\mathfrak{p}|^2 + C_3 \leqslant C_4 n^3 \end{aligned}$$

where $C_4 > 0$ is a constant independent of n. For all $\mathfrak{z} \in W[r]$ and all $n \in \mathbf{N}$, (7.7) and (7.9) imply

$$(7.10)\qquad \log|f_n(\mathfrak{z})| \leqslant C_4 n^3 + C_1(1 + r + n|\mathfrak{p}|)^\mu.$$

Here the constants $C_4 > 0$ and $C_1 > 0$ do not depend on $n \in \mathbf{N}, r \in \mathbf{R}_+$ and $\mathfrak{z} \in W[r]$.

According to Lemma 7.5 and considering (7.4), (7.6) and (7.10), there are constants $C_5 > 0$ and $C_6 > 0$ such that for all $n \in \mathbf{N}$ and $r \in \mathbf{R}_+$ and $\mathfrak{z} \in W[r/2]$ the following estimate holds:

$$(7.11)\qquad |A_n(\mathfrak{z})| \leqslant C_5 n^3 + C_6(1 + r + n|\mathfrak{p}|)^\mu.$$

Because Q is a homogeneous polynomial of degree 2 on W, a bilinear map $B: W \times W \to \mathbf{C}$ is defined by

$$(7.12)\qquad B(\mathfrak{z}, \mathfrak{w}) = Q(\mathfrak{z} + \mathfrak{w}) - Q(\mathfrak{z}) - Q(\mathfrak{w})$$

for all $(\mathfrak{z}, \mathfrak{w}) \in W \times W$. Then

$$(7.13)\qquad \begin{aligned} A_n(\mathfrak{z}) &= \sum_{\lambda=0}^{n-1} [g(\mathfrak{z} + \lambda\mathfrak{p}) - g(\lambda\mathfrak{p})] \\ &= \sum_{\lambda=0}^{n-1} [C + L(\mathfrak{z}) + \lambda L(\mathfrak{p}) + Q(\mathfrak{z}) + \lambda B(\mathfrak{z}, \mathfrak{p}) + \lambda^2 Q(\mathfrak{p})] \\ &\quad - \sum_{\lambda=0}^{n-1} [C + \lambda L(\mathfrak{p}) + \lambda^2 Q(\mathfrak{p})] \\ &= nL(\mathfrak{z}) + nQ(\mathfrak{z}) + \tfrac{1}{2}n(n-1)B(\mathfrak{z}, \mathfrak{p}). \end{aligned}$$

Take any $\mathfrak{y} \neq 0$ in W. Take any $n \in \mathbf{N}$. Define

$$(7.14)\qquad r = n|\mathfrak{y}| \quad\text{and}\quad \mathfrak{z} = \tfrac{1}{2}(n-1)\mathfrak{y}.$$

Then

$$|\mathfrak{z}| = \tfrac{1}{2}(n-1)|\mathfrak{y}| \leqslant n|\mathfrak{y}| = \tfrac{1}{2}r \quad\text{and}$$

$$\begin{aligned} A_n(\mathfrak{z}) &= \tfrac{1}{2}n(n-1)L(\mathfrak{y}) + \tfrac{1}{4}n(n-1)^2(Q(\mathfrak{y}) + B(\mathfrak{y}, \mathfrak{p})) \\ &= \tfrac{1}{4}n(n-1)(\,2L(\mathfrak{y}) + (n-1)Q(\mathfrak{y}+\mathfrak{p}) - (n-1)Q(\mathfrak{p})\,). \end{aligned}$$

Hence (7.11) implies

$$(7.15)\qquad \begin{aligned} &\tfrac{1}{4}n(n-1)^2|2L(\mathfrak{y})/(n-1) + Q(\mathfrak{y} + \mathfrak{p}) - Q(\mathfrak{p})| \\ &\qquad\leqslant C_5 n^3 + C_6(1 + n|\mathfrak{y}| + n|\mathfrak{p}|)^\mu \end{aligned}$$

where C_5 and C_6 do not depend on n and $\mathfrak{y}$. Divide (7.15) by n^3 and let $n \to \infty$. Then

$$|Q(\mathfrak{y} + \mathfrak{p}) - Q(\mathfrak{p})| \leqslant 4C_5$$

for all $\mathfrak{y} \in W$. Consequently, Q is constant. Also Q is a homogeneous polynomial of degree 2. Hence $Q \equiv 0$. q.e.d.

THEOREM 7.6 [48]. *Let $0 \leqslant \nu \not\equiv 0$ be a nonnegative, abelian divisor on W. Then there exists a theta function h on W such that $\mu_h = \nu$ and $\mathfrak{Q}(h) = \mathfrak{P}(\nu)$. Moreover, if $0 \notin \operatorname{supp} \nu = A \not\equiv \varnothing$ then $q = 2$ is the genus of ν. Let h be the canonical function of ν for the constant weight 2. Then h is a theta function on W with $\mu_h = \nu$ and $\mathfrak{Q}(h) = \mathfrak{P}(\nu)$. Particularly, if s is maximal such that $A(s) = \varnothing$ and if $\mathfrak{z} \in W(s)$, then*

$$\log h(\mathfrak{z}) = \int_A \nu(\mathfrak{y}) e((\mathfrak{z} | \mathfrak{y})/(\mathfrak{y} | \mathfrak{y}), 2)\, \omega^{m-1}(\mathfrak{y}).$$

PROOF. W.l.o.g. $0 \notin A$. Lemma 7.1 shows that $q = 2$ is the genus of ν. Let h be the canonical function of ν for the weight 2. Then $\operatorname{Ord} h = 2$. According to Lemma 7.5, $\mathfrak{Q}(h) = \mathfrak{P}(\mu_h) = \mathfrak{P}(\nu)$. Because $\mathfrak{P}(\nu)$ is a lattice, h is a theta function. q.e.d.

Obviously, h is the analogue to the Weierstrass σ-function. By differentiation, the analogue to the $\wp$-function can be obtained. A lattice $\mathfrak{L}$ is said to be discrete, if and only if an open neighborhood U of $0 \in W$ exists such that $U \cap \mathfrak{L} = \{0\}$.

THEOREM 7.7 [48]. *Let $\nu \geqslant 0$ be a nonnegative, abelian divisor on W. Define $A = \operatorname{supp} \nu$. Assume $0 \notin A \not\equiv \varnothing$. Let s be maximal such that $A(s) = \varnothing$. Assume that $\mathfrak{P}(\nu)$ is discrete. Take vectors $\mathfrak{a} \neq 0$ and $\mathfrak{b} \neq 0$ in W. Then one and only one meromorphic function $\wp_{\mathfrak{ab}}$ on W exists such that $\wp_{\mathfrak{ab}}$ is holomorphic on $W(s)$ and such that*

$$\wp_{\mathfrak{ab}}(\mathfrak{z}) = \int_A \nu(\mathfrak{y}) e''_m \left(\frac{(\mathfrak{z} | \mathfrak{y})}{(\mathfrak{y} | \mathfrak{y})}, 2 \right) \frac{(\mathfrak{a} | \mathfrak{y})\,(\mathfrak{b} | \mathfrak{y})}{|\mathfrak{y}|^4}\, \omega^{m-1}(\mathfrak{y}).$$

Moreover, abbreviate $h = \wp_{\mathfrak{ab}}$, then $A = \operatorname{supp} \mu_h^\infty$. If $\mathfrak{z} \in \mathfrak{R}(A)$, then $\delta_{\mu_h^\infty}(\mathfrak{z}) = 2$. Moreover, $\wp_{\mathfrak{ab}}$ is a meromorphic abelian function with $\mathfrak{P}(\wp_{\mathfrak{ab}}) = \mathfrak{P}(\nu)$.

In view of the properties of the $\wp$-functions in one variable, it would be of interest to know, which field is generated by these functions $\wp_{\mathfrak{ab}}$, which relation does exist between these functions $\wp_{\mathfrak{ab}}$ and their directional derivatives, and if the explicit integral representation of these functions can be used to obtain additional information in the theory of theta functions.

THEOREM 7.8 (APPELL [2], POINCARÉ [34]). *Let $f \not\equiv 0$ be a meromorphic, abelian function on W. Then there exist theta functions g and h on W such that $hf = g$, and such that g and h are coprime at every point of W. Also $\mathfrak{P}(f) \subseteq \mathfrak{Q}(g) \cap \mathfrak{Q}(h)$.*

PROOF. W.l.o.g. it can be assumed that f is holomorphic at $0 \in W$ and that $f(0) = 1$. Then μ_f^0 and μ_f^∞ are abelian divisors with $\mathfrak{P}(f) \subseteq \mathfrak{P}(\mu_f^0) \cap \mathfrak{P}(\mu_f^\infty)$. Let g_a be the canonical function to μ_f^a for the weight 2 for $a = 0$ and $a = \infty$. Then g_0 and g_∞ are theta functions with $\mu_{g_a} = \mu_f^a$ and $\mathfrak{P}(\mu_f^a) = \mathfrak{P}(\mu_{g_a}) = \mathfrak{Q}(g_a)$ for $a = 0$ and $a = \infty$. A holomorphic function $Q: W \to \mathbf{C}$ with $Q(0) = 0$ exists uniquely such that $f = e^Q g_0 / g_\infty$. Obviously $\mathfrak{Q}(Q) \supseteq \mathfrak{P}(f) \cap \mathfrak{Q}(g_0) \cap \mathfrak{Q}(g_\infty) = \mathfrak{P}(f)$. Hence e^Q is a theta function. Hence $g = e^Q g_0$ and $h = g_\infty$ are theta functions with $\mathfrak{Q}(g_0) = \mathfrak{Q}(g)$ and $\mathfrak{Q}(g_\infty) = \mathfrak{Q}(h)$. Hence $hf = g$ and $\mathfrak{P}(f) \subseteq \mathfrak{Q}(g) \cap \mathfrak{Q}(h)$. q.e.d.

8. Functions of Finite λ-Type

Kujala [23] extended the Fourier series methods of Rubel and Taylor [36] to several variables. Another extension was given by Taylor [61]. Here a short outline of results of Kujala [23] shall be given without proofs. For more details see the original papers.

Let $f \not\equiv 0$ be a meromorphic function on W. Assume that f is holomorphic at $0 \in W$ with $f(0) \neq 0$. Define the Fourier coefficients of $\log |f|$ by

$$C_p(r; f; \mathfrak{a}) = \frac{1}{2\pi} \int_0^{2\pi} \log |f(re^{it}\mathfrak{a})| e^{-ipt}\, dt$$

for all $r > 0$ and $p \in \mathbf{Z}$ and $\mathfrak{a} \in W$.

If $f \not\equiv 0$ and $g \not\equiv 0$ are meromorphic functions on W which are holomorphic at 0 with $f(0) \neq 0 \neq g(0)$, then

$$C_{-p}(r; f; \mathfrak{a}) = \overline{C_p(r; f; \mathfrak{a})},$$

$$C_p(r; fg; \mathfrak{a}) = C_p(r; f; \mathfrak{a}) + C_p(r; g; \mathfrak{a}),$$

$$C_p(r; f^{-1}; \mathfrak{a}) = -C_p(r; f; \mathfrak{a}),$$

$$C_p(r; f; z\mathfrak{a}) = C_p(r\,|z|; f; \mathfrak{a}) \quad \text{for } 0 \neq z \in \mathbf{C}.$$

Moreover, the Fourier series $\Sigma_{p=-\infty}^{+\infty} C_p(r; f; \mathfrak{a}) e^{ipt}$ converges in the L^2-norm on the L^2-functions on $\mathbf{R}[0, 2\pi]$ to the function $\log |f(re^{it}\mathfrak{a})|$.

Let f be a meromorphic function on W which is holomorphic at $0 \in W$ with $f(0) \neq 0$. Take $s > 0$ such that f is holomorphic on $W(s)$ and $f(\mathfrak{z}) \neq 0$ for all $\mathfrak{z} \in W(s)$. Define

$$f'(\mathfrak{z}) = \frac{d}{dx} f(x\mathfrak{z}) \Big|_{x=1}$$

for $\mathfrak{z} \in W$. Then

$$f'/f = \sum_{p=1}^{\infty} p\alpha_p(f)$$

on $W(s)$ where α_p is a homogeneous polynomial of degree p.

Let ν be a divisor on W with $0 \notin \operatorname{supp} \nu$. For $r > 0, p \in \mathbf{N}$ and $\mathfrak{a} \in W$ define

$$N'_p(r; \nu; \mathfrak{a}) = \frac{1}{p} \sum_{z \in \mathbf{C}[r]} \nu(z; \mathfrak{a})\, (r/z)^p,$$

$$N''_p(r; \nu; \mathfrak{a}) = \frac{1}{p} \sum_{z \in \mathbf{C}[r]} \nu(z; \mathfrak{a})\, (\bar{z}/r)^p,$$

$$N_p(r; \nu; \mathfrak{a}) = N'_p(r; \nu; \mathfrak{a}) - N''(r; \nu; \mathfrak{a}).$$

The $N_p(r; \nu; \mathfrak{a})$ is a continuous function of $(r; \mathfrak{a})$ on $\mathbf{R}^+ \times W$.

Assume that f is a meromorphic function on W which is holomorphic at $0 \in W$ with $f(0) \neq 0$. Let $\nu = \mu_f$ be the divisor of f. Take $r > 0$ and $\mathfrak{a} \in W$ and $p \in \mathbf{N}$. Then

$$C_0(r; f; \mathfrak{a}) = \log |f(0)| + N_\nu(r; \mathfrak{a}),$$
$$2C_p(r; f; \mathfrak{a}) = r^p \alpha_p(f)(\mathfrak{a}) + N_p(r; \nu; \mathfrak{a}),$$

which is the Fourier series development of the Jensen-Poisson formula.

A continuous, increasing function $\lambda : \mathbf{R}[0, \infty) \to \mathbf{R}(0, \infty)$ is said to be a growth function. Assume through this section that a growth function λ is given. A meromorphic function f is said to be of finite λ-type if there are constants $s > 0$ and $B > 0$ such that

$$\limsup_{r \to \infty} T_f(r, s)/\lambda(Br) < \infty.$$

The set $\mathfrak{M}(\lambda, W)$ of meromorphic functions of finite λ-type on W is a field. Let $\mathfrak{H}(\lambda, W)$ be the integral domain consisting of the holomorphic functions in $\mathfrak{M}(\lambda, W)$. The main problem is to determine when $\mathfrak{M}(\lambda, W)$ is the field of quotients of $\mathfrak{M}(\lambda, W)$.

Let f be a meromorphic function on W such that f is holomorphic at 0 with $f(0) \neq 0$. Then f is of finite λ-type if and only if there are positive constants A, B and r_0 such that $T_f(r; \mathfrak{a}) \leqslant A\lambda(Br)$ for all $r > r_0$ and all $\mathfrak{a} \in W\langle 1\rangle$.

Let ν be a nonnegative divisor on W. Then ν is of finite λ-density if and only if there are constants $s > 0$ and $B > 0$ such that

$$\limsup_{r \to \infty} N_f(\mathrm{r}, \mathrm{s})/\lambda(Br) < \infty.$$

If $0 \notin \operatorname{supp} \nu$, Proposition 3.9 implies that ν is of finite λ-density if and only if there are constants $A > 0, B > 0$ and $r_0 > 0$ such that $N(r; \mathfrak{a}) \leqslant A\lambda(Br)$ for all $r \geqslant r_0$ and all $\mathfrak{a} \in W\langle 1\rangle$.

Again, let ν be a nonnegative divisor on W with $0 \notin \operatorname{supp} \nu$. Then ν is said to be λ-balanced if and only if there are positive constants A, B and r_0 such that

$$\left| \frac{1}{p} \sum_{s < |z| \leqslant r} \nu(z; \mathfrak{a}) z^{-p} \right| \leqslant \frac{A\lambda(Br)}{r^p} + \frac{A\lambda(Bs)}{s^p}$$

for all $p \in \mathbf{N}$, all $\mathfrak{a} \in W\langle 1\rangle$, for all $r > s > r_0$. The divisor ν is said to be λ-admissible if and only if ν is λ-balanced and has finite λ-density.

LEMMA 8.1 (KUJALA [23]). *Let $\nu \geqslant 0$ be a nonnegative divisor on W with $0 \notin \operatorname{supp} \nu$. Assume that there exists a sequence $\{\alpha_p\}_{p \in \mathbf{N}}$ of functions $\alpha_p : W\langle 1\rangle \to \mathbf{C}$ and positive constants A, B and r_0 such that*

$$|r^p \alpha_p(\mathfrak{a}) + N_p(r; \nu; \mathfrak{a})| \leqslant A\lambda(Br)$$

for all $p \in \mathbf{N}$, all $\mathfrak{a} \in W\langle 1\rangle$ and all $r > r_0$. Then ν is λ-admissible.

LEMMA 8.2 (KUJALA [23]). *Let $\nu \geqslant 0$ be a nonnegative divisor on W with $0 \notin \operatorname{supp} \nu$. Assume that ν is λ-admissible. Then there exist a family $\{\alpha_p\}_{p \in \mathbf{N}}$ of continuous* *functions* $\alpha_p : W\langle 1\rangle \to \mathbf{C}$ *and positive constants* A, B, r_0 *such that*

$$|r^p \alpha_p(\mathfrak{a}) + N_p(r; \nu; \mathfrak{a})| \leqslant A\lambda(Br)$$

for all $r > r_0$, *all* $\mathfrak{a} \in W\langle 1\rangle$ *and all* $p \in \mathbf{N}$.

These lemmata are used in the proof of the main result.

THEOREM 8.3 (KUJALA [23]). *Let* $\nu \geqslant 0$ *be a nonnegative divisor on* W *with* $0 \notin \operatorname{supp} \nu$. *Then there exists a holomorphic function of finite* λ*-type on* W *with* $\nu = \mu_f$ *if and only if* ν *is* λ*-admissible.*

Now, answers to the quotient field theorem can be given.

THEOREM 8.4 (KUJALA [23]). *Assume that positive constants* A, B, r_0 *and* p_0 *exist such that*

$$\int_s^r \lambda(t) t^{-p-1}\, dt \leqslant A\lambda(Br) r^{-p} + A\lambda(Bs) s^{-p}$$

for all $r \geqslant s > r_0$ *and all* $p \geqslant p_0$. *Then* $\mathfrak{M}(\lambda, W)$ *is the field of quotients of* $\mathfrak{H}(\lambda, W)$.

Observe that the assumption on λ is satisfied, if there are constants $B > 1$ and $C > 0$ and $r_0 > 0$ such that $\lambda(Br) \leqslant C\lambda(r)$ for all $r \geqslant r_0$. Especially this is true for $\lambda_\rho(r) = 1 + r^\rho$ where $\rho > 0$ is a constant. For the case λ_ρ see also [56].

Similar results are obtained for functions of zero λ-type by Kujala [23].

9. Functions of Finite Order on the Unit Ball

Previously, functions of finite order on $W = W(\infty)$ were studied. Now, the case $R < \infty$ shall be considered. W.l.o.g. $R = 1$ can be assumed. This has been carried out in the Notre Dame thesis of Mueller [31] and in a different fashion by Kujala. Since Kujala's results are not available to me in writing, the results of [31] only shall be present here.

Let $s : \mathbf{R}[a, 1) \to \mathbf{R}_+$ be an increasing function with $0 \leqslant a < 1$. Define

$$0 \leqslant \operatorname{Ord} s = \limsup_{r \to 1} \frac{\log^+ s(r)}{-\log(1-r)} \leqslant +\infty$$

as the order of s.

If $\nu \geqslant 0$ is a nonnegative divisor on $W(1)$, then

$$\operatorname{Ord} \nu = \operatorname{Ord} N_\nu(\Box, s)$$

is independent of $0 < s < 1$ and called the order of ν. If $0 \notin \operatorname{supp} \nu$, then $s = 0$ is permitted. Then

$$\operatorname{Ord} \nu = \operatorname{Max}(0, (\operatorname{Ord} n_\nu) - 1).$$

The number $\mu \geqslant 0$ is said to be an exponent of convergence of the divisor $\nu \geqslant 0$ with $A = \operatorname{supp} \nu$ if and only if

$$\int_A \nu(1 - \tau_0)^{\mu+1} \omega^{m-1} < \infty \quad \text{if } m > 1,$$

$$\sum_{\mathfrak{z} \in A} \nu(\mathfrak{z})(1 - |\mathfrak{z}|)^{\mu+1} < \infty \quad \text{if } m = 1.$$

Otherwise μ is said to be an exponent of divergence of ν. If $\mu > \operatorname{Ord} \nu$, then μ is an exponent of convergence of ν. If $0 \leqslant \mu < \operatorname{Ord} \nu$, then μ is an exponent of divergence of ν. The smallest nonnegative integer q which is an exponent of convergence of ν is said to be the genus of ν.

Again, let $\nu \geqslant 0$ be a nonnegative divisor on $W(1)$. Then ν is said to satisfy a Blaschke condition if and only if $\mu = 0$ is an exponent of convergence or equivalently if

$$N_\nu(1, s) = \int_s^1 n_\nu(t) \frac{dt}{t} < \infty$$

for some $s \in \mathbf{R}(0, 1)$ and hence for all $s \in \mathbf{R}(0, 1)$. If $0 \notin \operatorname{supp} \nu$, then $s = 0$ is permitted.

If $m = 1$, then ν is the divisor of a bounded, holomorphic function on the unit disc, if and only if ν satisfies the Blaschke condition. If $m > 1$, then each divisor of a bounded holomorphic function satisfies a Blaschke condition; but there are easy examples of nonnegative divisors of $W(1)$ satisfying a Blaschke condition which are not the divisors of a bounded holomorphic function.

Let V be a hermitian vector space of dimension $n+1>0$. Let $f: W(1) \longrightarrow \mathbf{P}(V)$ be a meromorphic map. Then

$$\operatorname{Ord} f = \operatorname{Ord} T_f(\square, s)$$

is called the order f and does not depend on $s \in \mathbf{R}(0, 1)$. If f is holomorphic at $0 \in W$, then $s=0$ is permitted. Assume that $f(W(1)) \not\subseteq \ddot{E}[a]$ for all $a \in \mathbf{P}(V^*)$. Then $\operatorname{Ord} \mu_f^a \leqslant \operatorname{Ord} f$ according to the First Main Theorem.

THEOREM 9.1 (MUELLER [31]). *Let $f \not\equiv 0$ be a holomorphic function on $W(1)$. Then*

$$\operatorname{Ord} \mu_f \leqslant \operatorname{Ord} f \leqslant \operatorname{Ord} \log^+ M_f \leqslant (2m-1) + \operatorname{Ord} \mu_f.$$

If $m=1$, these bounds are sharp. The question if the bound is sharp for $m>1$ seems to be open.

Now consider the case $m=1$ first. Then $W=\mathbf{C}$. Define

$$b(z, y) = \frac{1 - z/y}{1 - z\bar{y}}$$

for $z \in \mathbf{C}$ and $0 \neq y \in \mathbf{C}$ if $z\bar{y} \neq 1$. Define $u = 1 - b$. For $q \in \mathbf{Z}_+$ define

$$b(z, y, q) = E(u(z, y), q)$$

for $z \in \mathbf{C}$ and $0 \neq y \in \mathbf{C}$ with $z\bar{y} \neq 1$.

THEOREM 9.2 (TSUIJ [64], MUELLER [31]). *Let $\nu \geqslant 0$ be a nonnegative divisor on $\mathbf{C}(1)$ with finite order λ. Let $q \in \mathbf{N}$ be an exponent of convergence ν. Then $q \geqslant \lambda$. Choose $\epsilon \geqslant 0$ such that $\epsilon = 0$ if $q = \lambda$ and $0 < \epsilon < 1 - (\lambda - [\lambda])$ if $q > \lambda$. Define $A = \operatorname{supp} \nu$. Assume $0 \notin A \neq \varnothing$. Define $s = \{r \mid A(r) = \varnothing\}$. Then the product*

$$B(z) = \prod_{y \in A} b(z, y, q)^{\nu(y)}$$

exists for all $z \in \mathbf{C}(1)$ and converges uniformly on each compact subset of $\mathbf{C}(1)$. The function $B: \mathbf{C}(1) \to \mathbf{C}$ is holomorphic with $\mu_B = \nu$ and $\operatorname{Ord} B = \operatorname{Ord} \nu$. A constant $K > 0$ depending on s and q only exists such that

$$\log^+ |B(z)| \leqslant K \sum_{y \in A} \nu(y) \left(\frac{1 - |y|^2}{|1 - z\bar{y}|} \right)^{\lambda + 1 + \epsilon}$$

for all $z \in W(1)$. Define $K' = (\lambda + 1 + \epsilon) 2^{\lambda + 1 + \epsilon} K$. If $0 < r < 1$, then

$$\log M_B(r) \leqslant K'(1-r)^{-\lambda - 1 - \epsilon} \int_0^1 n_\nu(t) (1-t)^{\lambda + \epsilon} \, dt.$$

Here B is said to be the canonical function of ν for q.

If $N_\nu(1) < \infty$, a similar statement holds with $\lambda = 0$, namely

$$B(z) = \prod_{y \in A} b(z, y)^{\nu(y)}$$

is a bounded holomorphic function with $\mu_B = \nu$ and $|Be^{-N_\nu(1)}| \leqslant 1$. Here $B \exp(-N_\nu(1))$ is the classical Balschke product.

Let $f \not\equiv 0$ be a meromorphic function on $\mathbf{C}(1)$ of finite order λ. Assume that f is holomorphic at 0 with $f(0) = 1$. For $a = 0$ or $a = \infty$ let q_a be the genus of the divisor μ_f^a. Then $q_a \leqslant \lambda + 1$. Let B_a be the canonical function of μ_f^a for q_a. Then

$$H = fB_\infty (B_0)^{-1} : \mathbf{C}(1) \longrightarrow \mathbf{C} - \{0\}$$

is a holomorphic function of finite order μ on $\mathbf{C}(1)$ without zeros. Let q be the smallest nonnegative integer such that

$$\int_0^1 (1 - r)^{q-1} T_H(r)\, dr < \infty \quad \text{if } q > 0,$$

$$T_H(1) < \infty \quad \text{if } q = 0.$$

Then $\mu \leqslant q \leqslant \mu + 1$. A function $\psi : \mathbf{C}\langle 1\rangle \to \mathbf{R}$ of bounded variation exists such that

$$\int_{\mathbf{C}(1)} d\psi = 0 \quad \text{and} \quad h(z) = \frac{1}{2\pi} \int_{\mathbf{C}\langle 1\rangle} \left(2\left(\frac{w}{w - z}\right)^{q+1} - 1 \right) d\psi(w)$$

where h is holomorphic with $h(0) = 0$ and $H = e^h$. Observe $f = e^h B_0 / B_\infty$ on $\mathbf{C}(1)$. Moreover $\lambda \leqslant \operatorname{Max}(\mu, q_0, q_\infty) \leqslant \lambda + 1$. This generalizes the well-known factorization in the case $T_f(1) < \infty$.

Now, the case of the unit ball $W(1)$ in a hermitian vector space W with $\dim W = m > 1$ shall be considered. For $0 \neq \mathfrak{y} \in W(1)$ and $\mathfrak{z} \in W(|\mathfrak{y}|)$ and $0 \leqslant q \in \mathbf{Z}$ define

$$b_m(\mathfrak{z}, \mathfrak{y}, q) = E\left(\frac{(\mathfrak{z}|\mathfrak{y})}{(\mathfrak{y}|\mathfrak{y})} \, \frac{1 - |\mathfrak{y}|^2}{1 - (\mathfrak{z}|\mathfrak{y})}, q \right)$$

$$\beta_m(\mathfrak{z}, \mathfrak{y}, q) = \frac{1}{(m-1)!} \frac{d^{m-1}}{dz^{m-1}} \left[z^{m-1} \log b_m(z\mathfrak{z}, \mathfrak{y}, q) \right] \Big|_{z=1}.$$

THEOREM 9.3 (MUELLER [31]). *Let $\nu \geqslant 0$ be a nonnegative divisor on $W(1)$. Define $A = \operatorname{supp} \nu$. Assume $0 \notin A \neq \varnothing$. Define $s = \sup \{r | A(r) = \varnothing\}$. Assume that $q \in \mathbf{Z}_+$ is an exponent of convergence of ν. Then there exists one and only one holomorphic function $B : W(1) \to \mathbf{C}$ with $B(0) = 1$ such that $B(\mathfrak{z}) \neq 0$ if $\mathfrak{z} \in W(s)$ and such that*

$$\log B(\mathfrak{z}) = \int_A \nu(\mathfrak{y}) \beta_m(\mathfrak{z}, \mathfrak{y}, q) \omega^{m-1}(\mathfrak{y})$$

for all $\mathfrak{z} \in W(s)$. Moreover this canonical function B of ν for q has the following properties:

(1) *The divisor ν is the zero divisor of B, that is $\mu_B = \nu$.*

(2) *If $\operatorname{Ord} \nu = \lambda$, take $\epsilon \geqslant 0$ such that $\epsilon = 0$ if $q = \lambda$ and $0 < \epsilon < 1 - (\lambda - [\lambda])$ if $q > \lambda$. Then there exists a constant $K > 0$ such that*

$$(m-1) \int_0^1 (1 - t)^{m-2} \log |B(t\mathfrak{z})|\, dt \leqslant K \int_A \nu(\mathfrak{y}) \left(\frac{1 - |\mathfrak{y}|^2}{|1 - (\mathfrak{z}|\mathfrak{y})|} \right)^{\lambda + 1 + \epsilon} \omega^{m-1}(\mathfrak{y})$$

for all $\mathfrak{z} \in W\langle 1\rangle$.

(3) *A constant $L > 0$ exists such that for all r and θ with $0 < \theta < 1$ and $0 < r < 1$ the following estimate holds:*

$$T_B(\theta^2 r) \leqslant \frac{2}{(1 - \theta)^{m-1}} \left(\frac{L}{(1 - r)^{\lambda + 1 + \epsilon}} + (2m - 1) n_\nu(r) + N_\nu(r) \right).$$

(4) *The following estimate holds*:

$$\operatorname{Ord} \nu \leqslant \operatorname{Ord} B \leqslant m + 2 + \operatorname{Ord} \nu.$$

It would be interesting to find out if this estimate is sharp or if it can be improved. Now, the analogue of the quotient representation of a meromorphic function shall be considered.

Let f be a meromorphic function of finite order λ on the unit ball $W(1)$ with $\dim W = m > 1$. Assume that f is holomorphic at $0 \in W$ with $f(0) = 1$. For $a = 0$ and $a = \infty$ let q_a be the genus of the divisor μ_f^a. Then $q_a \leqslant \lambda + 1$. Let B_a be the canonical function of μ_f^a for q_a. Holomorphic functions h and H with $h(0) = 0$ exist on $W(1)$ such that $H = e^h = fB_\infty/B_0$. The function H has finite order μ. A positive integer q is uniquely defined by

$$m + \mu < q \leqslant m + \mu + 1.$$

Then a continuous function $L\colon W[1] \to \mathbf{C}$ is defined by

$$L(\mathfrak{z}) = q \int_0^1 (1 - t)^{q-1} h(t\mathfrak{z})\, dt$$

for $\mathfrak{z} \in W[1]$. Moreover $L \mid W(1)$ is holomorphic. Define $\psi = \tfrac{1}{2}(L + \bar{L})$. Then

$$h(\mathfrak{z}) = \int_{W\langle 1\rangle} 2 \sum_{\alpha=0}^{q} \binom{q}{\alpha} \binom{m + \alpha - 1}{\alpha} \left(\frac{(\mathfrak{z} \mid \mathfrak{y})^\alpha}{(1 - (\mathfrak{z} \mid \mathfrak{y}))^{m+\alpha}} - 1 \right) \psi(\mathfrak{y})\sigma(\mathfrak{y})$$

for all $\mathfrak{z} \in W(1)$. Moreover

$$\max(\mu, q_0, q_\infty) \leqslant \lambda + p + 2, \qquad q \leqslant \lambda + 2p + 3.$$

It is not known if these estimates are sharp.

10. **Functions of Finite Order on Polydiscs**

Ronkin [35] has given another representation of the canonical function of a nonnegative divisor on W which is adapted to an exhaustion by polydiscs of W. Ronkin uses a power series method which closely resembles the original method of Kneser and the Fourier series method of Rubel-Taylor-Kujala. Here, a different method will be used. The whole theory will be recast, reorganized and considerably extended. Since this seems to be new, complete proofs will be provided in this section.

Unfortunately, quite a number of new notations will be needed. Recall

$$\begin{aligned} \mathbf{R}^+ &= \mathbf{R}(0, \infty) = \{x \in \mathbf{R} \mid x > 0\}, \\ \mathbf{R}^{+m} &= \mathbf{R}^+ \times \cdots \times \mathbf{R}^+ \quad (m \text{ times}), \\ \mathbf{R}_+ &= \mathbf{R}[0, \infty) = \{x \in \mathbf{R} \mid x \geqslant 0\}. \end{aligned}$$

Define

$$\begin{aligned} \mathbf{C}_* &= \mathbf{C} - \{0\}, \\ \mathbf{C}_*^m &= \mathbf{C}_* \times \cdots \times \mathbf{C}_* \quad (m \text{ times}), \\ {}^*\mathbf{P}_{m-1} &= \mathbf{P}(\mathbf{C}_*^m) \subseteq \mathbf{P}_{m-1} = \mathbf{P}(\mathbf{C}^m). \end{aligned}$$

Observe $\mathbf{R}^{+m} \subseteq \mathbf{C}_*^m \subseteq \mathbf{C}^m$. For $\mathfrak{z} = (z_1, \cdots, z_m) \in \mathbf{C}^m$ define

$$\|\mathfrak{z}\| = \max\,(|z_1|, \cdots, |z_m|).$$

For each $\mathfrak{x} = (x_1, \cdots, x_m) \in \mathbf{R}^{+m}$ and $r \in \mathbf{R}^+$ define

$$T_{\mathfrak{x}}\langle \cdot r \cdot \rangle = \prod_{\mu=1}^{m} \mathbf{C}\langle r x_\mu \rangle,$$

$$P_{\mathfrak{x}}(\cdot r \cdot) = \prod_{\mu=1}^{m} \mathbf{C}(r x_\mu), \qquad P_{\mathfrak{x}}[\cdot r \cdot] = \prod_{\mu=1}^{m} \mathbf{C}[r x_\mu],$$

$$\begin{aligned} P_{\mathfrak{x}} &= P_{\mathfrak{x}}(\cdot 1 \cdot), \qquad T_{\mathfrak{x}} = T_{\mathfrak{x}}\langle \cdot 1 \cdot \rangle, \\ \widetilde{T}_{\mathfrak{x}} &= \{z\mathfrak{y} \mid z \in \mathbf{C} \text{ and } \mathfrak{y} \in T_{\mathfrak{x}}\}, \\ \ddot{T}_{\mathfrak{x}} &= \mathbf{P}(T_{\mathfrak{x}}) = \mathbf{P}(\widetilde{T}_{\mathfrak{x}}) \subseteq {}^*\mathbf{P}_{m-1}, \\ T_{\mathfrak{x}}^{\nu} &= \{(z_1, \cdots, z_m) \in T_{\mathfrak{x}} \mid z_\nu = x_\nu\}. \end{aligned}$$

If $A \subseteq \mathbf{C}^m$, if $r \in \mathbf{R}^+$ and if $\mathfrak{x} \in \mathbf{R}^{+m}$, define

$$A\{\mathfrak{x}\} = A \cap \widetilde{T}_{\mathfrak{x}}, \qquad A(\mathfrak{x}; r) = A \cap \widetilde{T}_{\mathfrak{x}} \cap \mathbf{C}^m(r),$$

$$A[\mathfrak{x}; r] = A \cap \widetilde{T}_{\mathfrak{x}} \cap \mathbf{C}^m[r], \qquad A\langle \mathfrak{x}, r\rangle = A \cap \widetilde{T}_{\mathfrak{x}} \cap \mathbf{C}^m\langle r\rangle.$$

If $|\mathfrak{x}| = 1$, then

$$A(\mathfrak{x}, r) = \{z\mathfrak{y} \in A \mid \mathfrak{y} \in T_{\mathfrak{x}} \text{ and } z \in \mathbf{C}(r)\},$$

$$A[\mathfrak{x}, r] = \{z\mathfrak{y} \in A \mid \mathfrak{y} \in T_{\mathfrak{x}} \text{ and } z \in \mathbf{C}[r]\},$$

$$A\langle \mathfrak{x}, r\rangle = \{z\mathfrak{y} \in A \mid \mathfrak{y} \in T_{\mathfrak{x}} \text{ and } z \in \mathbf{C}\langle r\rangle\}.$$

Here $P_{\mathfrak{x}}(\cdot r \cdot)$ is a polydisc and $T_{\mathfrak{x}}\langle \cdot r \cdot\rangle$ is the distinguished boundary of this polydisc. Here $T_{\mathfrak{x}}\langle \cdot r \cdot\rangle$ is a real torus of dimension m. For each $\nu \in \mathbf{N}[1, m]$, the set $T_{\mathfrak{x}}^{\nu}$ is a real torus of dimension $m - 1$. Hence $T_{\mathfrak{x}}\langle \cdot r \cdot\rangle$ and $T_{\mathfrak{x}}^{\nu}$ are oriented, compact, connected manifolds. The restriction $\mathbf{P} : T_{\mathfrak{x}}^{\nu} \to \ddot{T}_{\mathfrak{x}}$ is bijective. Hence, one and only one oriented, real analytic structure exists on $\ddot{T}_{\mathfrak{x}}$ such that $\mathbf{P} : T_{\mathfrak{x}}^{m} \to \ddot{T}_{\mathfrak{x}}$ is a diffeomorphism. Here diffeomorphism means an orientation preserving diffeomorphism of class C^{∞} and it will be used in this sense.

Define $\widetilde{\gamma} : \mathbf{R}^{m-1} \to T_{\mathfrak{x}}$ by

$$\widetilde{\gamma}(\phi_1, \cdots, \phi_{m-1}) = (x_1 e^{i\phi_1}, \cdots, x_{m-1} e^{i\phi_{m-1}}, x_m).$$

Define

$$\gamma = \mathbf{P} \circ \widetilde{\gamma} : \mathbf{R}^{m-1} \longrightarrow \ddot{T}_{\mathfrak{x}}.$$

Define $\gamma_0 = \gamma \mid \mathbf{R}(0, 2\pi)^{m-1}$. Then $\gamma_0 : \mathbf{R}(0, 2\pi)^{m-1} \to \operatorname{Im} \gamma_0$ is a diffeomorphism onto an open subset of $\ddot{T}_{\mathfrak{x}}$ such that $\ddot{T}_{\mathfrak{x}} - \operatorname{Im} \gamma_0$ has measure zero.

Define $\alpha : \mathbf{R}^m \to T_{\mathfrak{x}}$ by

$$\alpha(\phi_1, \cdots, \phi_m) = (x_1 e^{i\phi_1}, \cdots, x_m e^{i\phi_m}).$$

Define $\alpha_0 = \alpha \mid \mathbf{R}(0, 2\pi)^m$. Then $\alpha_0 : \mathbf{R}(0, 2\pi)^m \to \operatorname{Im} \alpha_0$ is a diffeomorphism onto an open subset of $T_{\mathfrak{x}}$ and $T_{\mathfrak{x}} - \operatorname{Im} \alpha_0$ is a set of measure zero on $T_{\mathfrak{x}}$.

On $\mathbf{C}_*^m$ define the holomorphic differential form Ω by

$$\Omega(z_1, \cdots, z_m) = \left(\frac{1}{2\pi i}\right)^m \frac{dz_1}{z_1} \wedge \cdots \wedge \frac{dz_m}{z_m}.$$

Take $\mathfrak{x} \in \mathbf{R}^{+m}$. Let $j : T_{\mathfrak{x}} \to \mathbf{C}_*^m$ be the inclusion map. Then $j^*(\Omega) > 0$ is a volume form on $T_{\mathfrak{x}}$ with

$$\alpha^* j^*(\Omega) = \left(\frac{1}{2\pi}\right)^m d\phi_1 \wedge \cdots \wedge d\phi_m.$$

Therefore $\int_{T_{\mathfrak{x}}} \Omega = 1$.

Take ν and μ in $\mathbf{N}[1, m]$. A meromorphic function ζ_{μ}^{ν} exists uniquely on $\mathbf{P}_{m-1}$ such that ζ_{μ}^{ν} is holomorphic on ${}^*\mathbf{P}_{m-1}$ and

$$\zeta_{\mu}^{\nu} \circ \mathbf{P}(z_1, \cdots, z_m) = z_{\nu}/z_{\mu}.$$

Then $\zeta_{\mu}^{\nu} \cdot \zeta_{\lambda}^{\mu} = \zeta_{\lambda}^{\nu}$ and $\zeta_{\mu}^{\nu} \zeta_{\nu}^{\mu} = 1$ and $\zeta_{\nu}^{\nu} = 1$. A meromorphic differential is defined on $\mathbf{P}_{m-1}$ by

$$\eta_{\mu}^{\nu} = d\zeta_{\mu}^{\nu} / \zeta_{\mu}^{\nu}.$$

Then η_{μ}^{ν} is holomorphic on ${}^*\mathbf{P}_{m-1}$. Then

$$d\eta_\mu^\nu = 0, \qquad \eta_\nu^\nu = 0,$$
$$\eta_\mu^\nu + \eta_\lambda^\mu = \eta_\lambda^\nu, \qquad \eta_\mu^\nu + \eta_\nu^\mu = 0.$$

These identities show easily that the meromorphic differential

$$\ddot{\xi} = (-1)^{\nu-1}\left(\frac{i}{2\pi}\right)^{m-1} \eta_\nu^1 \wedge \cdots \wedge \eta_\nu^{\nu-1} \wedge \eta_\nu^{\nu+1} \wedge \cdots \wedge \eta_\nu^m$$

is independent of $\nu \in \mathbf{N}[1, m]$. The differential is holomorphic on ${}^*\mathbf{P}_{m-1}$. Observe that $\ddot{\xi}$ has bidegree $(m-1, 0)$ and that $d\ddot{\xi} = 0$. Then $\xi = \mathbf{P}^*(\ddot{\xi})$ is a holomorphic differential of bidegree $(m-1, 0)$ on $\mathbf{C}_*^m$ with $d\xi = 0$. An easy computation shows that

$$\xi \wedge dz_\nu / 2\pi i z_\nu = \Omega$$

for each $\nu \in \mathbf{N}[1, m]$.

Take $\mathfrak{r} \in \mathbf{R}^{+m}$. Let $j : \ddot{T}_\mathfrak{r} \to {}^*\mathbf{P}_{m-1}$ be the inclusion map. Then $j^*(\ddot{\xi}) > 0$ is a volume form on $\ddot{T}_\mathfrak{r}$ with

$$\gamma^* j^*(\ddot{\xi}) = \frac{1}{(2\pi)^{m-1}}\, d\phi_1 \wedge \cdots \wedge d\phi_{m-1}.$$

Therefore $\int_{\ddot{T}_\mathfrak{r}} \ddot{\xi} = 1$.

LEMMA 10.1. *Take* $\mathfrak{r} \in \mathbf{R}^{+m}$. *Take* $r \in \mathbf{R}^+$. *Let* $F : \ddot{T}_\mathfrak{r} \to \mathbf{C}$ *be a measurable function on* $\ddot{T}_\mathfrak{r}$. *Assume that at least one of the following two integrals exist. Then both exist and*

$$\int_{T_\mathfrak{r}\langle \cdot r \cdot \rangle} (F \circ \mathbf{P})\Omega = \int_{\ddot{T}} F\ddot{\xi}.$$

PROOF. A diffeomorphism $\mu_r : T_\mathfrak{r} \to T_\mathfrak{r}\langle \cdot r \cdot \rangle$ is defined by $\mu_r(\mathfrak{z}) = r\mathfrak{z}$. Then $F \circ \mathbf{P} \circ \mu_r = F \circ \mathbf{P}$. Let $j : T_\mathfrak{r} \to \mathbf{C}_*^m$ and $j_r : T_\mathfrak{r}\langle \cdot r \cdot \rangle \to \mathbf{C}_*^m$ be the inclusion maps. Then

$$j^*(\Omega) = \mu_r^* j_r^*(\Omega).$$

Hence it suffices to prove the lemma in the case $r = 1$. Define $\rho : \mathbf{R}(0, 2\pi)^m \to \mathbf{R}(0, 2\pi)^{m-1}$ by

$$\rho(\phi_1, \cdots, \phi_m) = (\phi_1 - \phi_m, \cdots, \phi_{m-1} - \phi_m).$$

Then

$$\mathbf{P} \circ \alpha = \gamma \circ \rho : \mathbf{R}^{m-1} \longrightarrow \ddot{T}_\mathfrak{r}.$$

The function $F \circ \gamma$ is periodic in each variable with a period 2π. Hence

$$\begin{aligned}
\int_{T_\mathfrak{r}} F \circ \mathbf{P}\Omega &= \left(\frac{1}{2\pi}\right)^m \int_0^{2\pi} \cdots \int_0^{2\pi} F \circ \mathbf{P} \circ \alpha(\phi_1, \cdots, \phi_m)\, d\phi_1 \cdots d\phi_m \\
&= \left(\frac{1}{2\pi}\right)^m \int_0^{2\pi} \cdots \int_0^{2\pi} F \circ \gamma \circ \rho(\phi_1, \cdots, \phi_m)\, d\phi_1 \cdots d\phi_m \\
&= \left(\frac{1}{2\pi}\right)^m \int_0^{2\pi} \cdots \int_0^{2\pi} F \circ \gamma(\phi_1, \cdots, \phi_{m-1})\, d\phi_1 \cdots d\phi_m \\
&= \left(\frac{1}{2\pi}\right)^{m-1} \int_0^{2\pi} \cdots \int_0^{2\pi} F \circ \gamma(\phi_1, \cdots, \phi_{m-1})\, d\phi_1 \cdots d\phi_{m-1} \\
&= \int_{\ddot{T}_\mathfrak{r}} F\ddot{\xi}. \quad \text{q.e.d.}
\end{aligned}$$

LEMMA 10.2. *Take* $r \in \mathbf{R}^+$ *and* $\mathfrak{x} \in \mathbf{R}^{+m}$. *Let* $F : T_{\mathfrak{x}}\langle \cdot r \cdot \rangle$ *be a function. Assume that* $F\Omega$ *is integrable over* $T_{\mathfrak{x}}\langle \cdot r \cdot \rangle$. *Then*

$$\int_{T_{\mathfrak{x}}\langle \cdot r \cdot \rangle} F\Omega = \int_{T_{\mathfrak{x}}} \frac{1}{2\pi} \int_0^{2\pi} F(re^{i\phi}\mathfrak{a})\, d\phi\, \Omega(\mathfrak{a})$$

where the inferior integral exists for almost all $\mathfrak{a} \in T_{\mathfrak{x}}$.

PROOF. W.l.o.g. $F \geqslant 0$ can be assumed. Define $H = \mathbf{R}(0, 2\pi) \times T_{\mathfrak{x}}\langle \cdot r \cdot \rangle$. Define $G : H \to \mathbf{R}_+$ by

$$G(\phi; \mathfrak{a}) = F(e^{i\phi}\mathfrak{a}).$$

Then G is measurable on H. Let $\pi : H \to T_{\mathfrak{x}}\langle \cdot r \cdot \rangle$ be the projection. Observe

$$\int_H G d\phi \wedge \pi^*(\Omega) = \int_0^{2\pi} \int_{T_{\mathfrak{x}}\langle \cdot r \cdot \rangle} F(e^{i\phi}\mathfrak{a})\Omega(\mathfrak{a})\, d\phi$$

$$= 2\pi \int_{T_{\mathfrak{x}}\langle \cdot r \cdot \rangle} F\Omega < +\infty.$$

Hence

$$\int_H G d\phi \wedge \pi^*\Omega = \int_{T_{\mathfrak{x}}\langle \cdot r \cdot \rangle} \int_0^{2\pi} F(e^{i\phi}\mathfrak{a})\, d\phi\, \Omega(\mathfrak{a})$$

$$= \int_{T_{\mathfrak{x}}} \frac{1}{2\pi} \int_0^{2\pi} F(e^{i\phi} r\mathfrak{a})\, d\phi\, \Omega(\mathfrak{a})$$

by virtue of the diffeomorphism μ_r of the proof of Lemma 10.1. q.e.d.

LEMMA 10.3. *Let* $F : \mathbf{C}^m - \{0\} \to \mathbf{C}$ *be a function. Assume that* $F\sigma$ *is integrable over* $\mathbf{C}^m\langle 1\rangle$. *Assume that* $F(z\mathfrak{a}) = F(\mathfrak{a})$ *for all* $z \in \mathbf{C}_*$ *and for all* $\mathfrak{a} \in W$. *Then*

$$H(\mathfrak{x}) = \int_{T_{\mathfrak{x}}} F\Omega$$

exists for almost all $\mathfrak{x} \in \mathbf{R}^{+m}$ *and*

$$\int_{\mathbf{C}^m\langle 1\rangle} F\sigma = 2^m \int_0^\infty \cdots \int_0^\infty He^{-\tau} x_1 \cdots x_m\, dx_1 \cdots dx_m$$

where $H = H(x_1, \cdots, x_m)$ *and* $\tau = x_1^2 + \cdots + x_m^2$.

PROOF. Lemma 2.1 implies

$$\int_{\mathbf{C}^m\langle 1\rangle} F\sigma = \frac{1}{m!} \int_{\mathbf{C}^m} Fe^{-\tau} \upsilon^m$$

$$= \int_{\mathbf{C}^m} Fe^{-\tau} \left(\frac{i}{2\pi}\right)^m dz_1 \wedge d\bar{z}_1 \wedge \cdots \wedge dz_m \wedge d\bar{z}_m$$

$$= 2^m \int_0^\infty \cdots \int_0^\infty He^{-\tau} x_1 \cdots x_m\, dx_1 \cdots dx_m$$

where

$$H(x_1, \cdots, x_m) = \left(\frac{1}{2\pi}\right)^m \int_0^{2\pi} \cdots \int_0^{2\pi} F(x_1 e^{i\phi_1} \cdots x_m e^{i\phi_m})\, d\phi_1 \cdots d\phi_m$$

$$= \int_{T_{\mathfrak{x}}} F\Omega$$

exists for almost all $\mathfrak{x} = (x_1, \cdots, x_m) \in \mathbf{R}^{+m}$. q.e.d.

LEMMA 10.4 *Take* $\mathfrak{x} \in \mathbf{R}^{+m}$ *with* $m > 1$. *Take* $0 < R \leqslant +\infty$. *Let* A *be a pure*

$(m-1)$-dimensional analytic subset of $P_{\mathfrak{x}}(\cdot R \cdot)$ with $0 \notin A$. Then the following statements hold:

(1) *The set $A\{\mathfrak{x}\}$ is real analytic in $P_{\mathfrak{x}}(\cdot R \cdot)$ and has pure dimension $m-1$.*

(2) *If N is a thin analytic subset of A, then $N\{\mathfrak{x}\}$ is nowhere dense in A.*

(3) *Let S be the set of all points $\mathfrak{z} \in A$ such that $\mathbf{P}: A \to \mathbf{P}_{m-1}$ is not locally biholomorphic at $\mathfrak{z}$. Then S is a thin analytic subset of A.*

(4) *Let $\mathfrak{R}(A\{\mathfrak{x}\})$ be the manifold of all simple points of $A\{\mathfrak{x}\}$. Then $A\{\mathfrak{x}\} - S$ is contained, open and dense in $\mathfrak{R}(A\{\mathfrak{x}\})$. Moreover, $\mathfrak{R}(A\{\mathfrak{x}\})$ carries one and only one orientation such that $\mathbf{P}: A\{\mathfrak{x}\} - S \to \ddot{T}_{\mathfrak{x}}$ is a local diffeomorphism. Affix this orientation to $\mathfrak{R}(A\{\mathfrak{x}\})$.*

(5) *The set $\mathbf{P}(S\{\mathfrak{x}\})$ has measure zero on $\ddot{T}$.*

(6) *Let $\iota: A\{\mathfrak{x}\} - S \to \mathbf{C}^m_*$ be the inclusion, then $\iota^*(\xi) > 0$ on $A\{\mathfrak{x}\} - S$ and $\iota^*(\xi) \geq 0$ on $A\{\mathfrak{x}\}$.*

PROOF. Because A is analytic in $P_{\mathfrak{x}}(\cdot R \cdot)$ and because $\widetilde{T}_{\mathfrak{x}}$ is real analytic in $\mathbf{C}^m$, the intersection $A\{\mathfrak{x}\} = A \cap \widetilde{T}_{\mathfrak{x}}$ is real analytic in $P_{\mathfrak{x}}(\cdot R \cdot)$. Because $0 \notin A$, the map $\mathbf{P}: A \to \mathbf{P}_{m-1}$ is light.

Define a biholomorphic map $\widetilde{\beta}: \mathbf{C}^{m-1} \times \mathbf{C}^* \to \mathbf{C}^{m-1} \times \mathbf{C}^*$ by $\widetilde{\beta}(\mathfrak{v}, u) = (u\mathfrak{v}, u)$ for all $\mathfrak{v} \in \mathbf{C}^{m-1}$ and $u \in \mathbf{C}^*$. Let $\pi: \mathbf{C}^{m-1} \times \mathbf{C}^* \to \mathbf{C}^{m-1}$ be the projection. Then one and only one biholomorphic map $\beta: \mathbf{C}^{m-1} \to Y$ onto an open, dense subset Y of $\mathbf{P}_{m-1}$ exists such that $\beta \circ \pi = \widetilde{\beta} \circ \mathbf{P}$. Identify $Y = \mathbf{C}^{m-1}$ such that β becomes the identity. Hence $\pi = \widetilde{\beta} \circ \mathbf{P}$.

Now, let N be a thin analytic subset of A. Assume that $N\{\mathfrak{x}\}$ is somewhere dense in $A\{\mathfrak{x}\}$. Hence an open subset U of $\mathbf{C}^m$ exists such that $N\{\mathfrak{x}\} \cap U = A\{\mathfrak{x}\} \cap U \neq \varnothing$. W.l.o.g $U \subseteq \mathbf{C}^{m-1} \times \mathbf{C}^*$ can be assumed. Take $\mathfrak{z}_0 \in N\{\mathfrak{x}\} \cap U$. Because $\mathbf{P}$ is light on A and because A has pure dimension $m-1$, an open connected neighborhood Q of $\mathfrak{z}_0$ in U and an open ball V in $\mathbf{C}^{m-1}$ centered at $\mathbf{P}(\mathfrak{z})$ exists such that $\mathbf{P}: Q \cap A \to V$ is light, proper and surjective. Therefore $N' = \mathbf{P}(Q \cap N)$ is analytic in V with $\dim N' \leq \dim N \leq m-2$. Hence $N' \neq V$. Therefore a holomorphic function $g \not\equiv 0$ on V exists such that $g(\mathfrak{v}) = 0$ for all $\mathfrak{v} \in N'$.

Now $\mathfrak{x} = (x_1, \cdots, x_m)$. Define

$$V_0 = V \cap \prod_{\mu=1}^{m-1} \mathbf{C}\langle x_\mu / x_m \rangle.$$

Take $\mathfrak{z} \in Q \cap N\{\mathfrak{x}\}$. Then $\mathfrak{z} = \widetilde{\beta}(\mathfrak{v}, u)$ with $\mathfrak{v} = (v_1, \cdots, v_m) \in \mathbf{C}^{m-1}$ and $u \in \mathbf{C}^*$. Then $\mathbf{P}(\mathfrak{z}) = \mathbf{P}(\widetilde{\beta}(\mathfrak{v}, u)) = \pi(\mathfrak{v}, u) = \mathfrak{v} \in V$. Also $\mathfrak{z} = (u\mathfrak{v}, u)$. Hence $|v_\mu| = x_\mu x_m^{-1}$. Therefore $\mathbf{P}(\mathfrak{z}) \in V_0$. Hence $\mathbf{P}(Q \cap N\{\mathfrak{x}\}) \subseteq V_0$ especially $V_0 \neq \varnothing$. Take $\mathfrak{v} = (v_1, \cdots, v_{m-1}) \in V_0$. Hence $\mathfrak{z} \in Q \cap A$ exists with $\mathbf{P}(\mathfrak{x}) = \mathfrak{v}$. Also $\pi\widetilde{\beta}^{-1}(\mathfrak{z}) = \mathbf{P}(\mathfrak{z}) = \mathfrak{v}$. Hence $\widetilde{\beta}^{-1}(\mathfrak{z}) = (\mathfrak{v}, u)$ with $u \in \mathbf{C}^*$ and $\mathfrak{z} = (u\mathfrak{v}, u) = (z_1, \cdots, z_m)$. Define $\mathfrak{y} = (x_m\mathfrak{v}, x_m)$. Then $|x_m v_\mu| = x_\mu$ for $\mu = 1, \cdots, m-1$. Hence $\mathfrak{y} \in T_{\mathfrak{x}}$. Therefore $\mathfrak{z} = u x_m^{-1} \mathfrak{y} \in \widetilde{T}_{\mathfrak{x}}$. Hence $\mathfrak{z} \in Q \cap A\{\mathfrak{x}\} = Q \cap N\{\mathfrak{x}\}$ which implies $\mathfrak{v} = \mathbf{P}(\mathfrak{z}) \in \mathbf{P}(Q \cap N\{\mathfrak{x}\})$. Therefore

$\mathbf{P}(Q \cap N\{\mathfrak{r}\}) = V_0$. Especially $N' \supseteq V_0$. Therefore $g \mid V_0 \equiv 0$. Because V_0 is an open, nonempty subset of a torus, $g \equiv 0$ on V. This contradiction shows that $N\{\mathfrak{r}\}$ is nowhere dense in $A\{\mathfrak{r}\}$.

Because A has pure dimension $m-1$ and because $\mathbf{P}: A \to \mathbf{P}_{m-1}$ is light, the set S defined in (3) is thin analytic in A. Especially, $\mathbf{P}: A - S \to \mathbf{P}_{m-1}$ is locally biholomorphic. Hence

$$A\{\mathfrak{r}\} - S = A\{\mathfrak{r}\} - S\{\mathfrak{r}\} = (A - S) \cap \mathbf{P}^{-1}(\ddot{T}_{\mathfrak{r}})$$

is contained and open in $\mathfrak{R}(A\{\mathfrak{r}\})$. Because $S\{\mathfrak{r}\}$ is nowhere dense in $A\{\mathfrak{r}\}$, the set $A\{\mathfrak{r}\} - S$ is dense in $\mathfrak{R}(A\{\mathfrak{r}\})$. Again, because $\mathbf{P}: A - S \to \mathbf{P}_{m-1}$ is locally biholomorphic, one and only one orientation can be affixed to $\mathfrak{R}(A\{\mathfrak{r}\})$ such that $\mathbf{P}: A\{\mathfrak{r}\} - S \to \ddot{T}_{\mathfrak{r}}$ is local diffeomorphic. Especially, the real analytic set $A\{\mathfrak{r}\}$ has pure real dimension $m-1$.

Observe that the $\ddot{T}_{\mathfrak{r}}$ is the distinguished boundary of some polydisc in $\mathbf{C}^{m-1} \subseteq \mathbf{P}_{m-1}$. Hence any almost thin subset in $\mathbf{P}_{m-1}$ intersects $\ddot{T}_{\mathfrak{r}}$ in a set of measure zero on $\ddot{T}_{\mathfrak{r}}$. Now, $\mathbf{P}(S)$ is almost thin in $\mathbf{P}_{m-1}$. Hence

$$\mathbf{P}(S\{\mathfrak{r}\}) = \mathbf{P}(S) \cap \ddot{T}_{\mathfrak{r}}$$

has measure zero on $\ddot{T}_{\mathfrak{r}}$.

Let $\iota: A\{\mathfrak{r}\} - S \to \mathbf{C}^m_*$ and $j: \ddot{T}_{\mathfrak{r}} \to {}^*\mathbf{P}_{m-1}$ be the inclusion maps. Then $\mathbf{P} \circ \iota = j \circ \mathbf{P}$. Since $j^*(\xi) > 0$, this implies $\mathbf{P}^* j^*(\ddot{\xi}) > 0$ on $A\{\mathfrak{r}\} - S$. Therefore

$$\iota^*(\xi) = \iota^*\mathbf{P}^*(\ddot{\xi}) = \mathbf{P}^* j^*(\ddot{\xi}) > 0 \quad \text{on } A\{\mathfrak{r}\} - S$$

by continuity $\iota^*(\xi) \geqslant 0$ on $\mathfrak{R}(A\{\mathfrak{r}\})$. q.e.d.

The statements of Lemma 10.4 are more or less explicitly contained in Ronkin [35]; however, no proofs are given there. The Lemma 10.1 can be extended to the case $0 \in A$ if some precautions are taken, but it would be too complicated to go into this here.

LEMMA 10.5. *Take* $0 < R \leqslant +\infty$. *Take* $\mathfrak{r} \in \mathbf{R}^{+m}$ *with* $|\mathfrak{r}| = 1$. *Let* $\nu \geqslant 0$ *be a nonnegative divisor on* $P_{\mathfrak{r}}(\cdot R \cdot)$. *Define* $A = \operatorname{supp} \nu$. *Assume* $0 \notin A \neq \emptyset$. *Let* $g: A\{\mathfrak{r}\} \to \mathbf{C}$ *be a function such that* $g\xi$ *is measurable on* $A\{\mathfrak{r}\}$. *Assume that either*

(a) *the form* $\nu g \xi$ *is integrable over* $A\{\mathfrak{r}\}$ *or*

(b) *the infinite series* $\Sigma_{|u|<R}\, \nu(u; \mathfrak{a}) |g(u\mathfrak{a})| \Omega(\mathfrak{a})$ *converges for almost all* $\mathfrak{a} \in T_{\mathfrak{r}}$ *and is integrable over* $T_{\mathfrak{r}}$.

Then (a) *and* (b) *are true. The series*

$$L(\mathfrak{a}) = \sum_{|u|<R} \nu(u; \mathfrak{a}) g(u\mathfrak{a})$$

converges absolutely almost everywhere on $T_{\mathfrak{r}}$. *A function* ${}^{\bullet}L^{\bullet}$ *exists almost everywhere on* $\ddot{T}_{\mathfrak{r}}$ *such that* ${}^{\bullet}L^{\bullet} \circ \mathbf{P} = L$ *almost everywhere on* $T_{\mathfrak{r}}$. *Then*

$$\int_{A\{\mathfrak{r}\}} \nu g \xi = \int_{\ddot{T}} {}^{\bullet}L^{\bullet} \ddot{\xi} = \int_{T_{\mathfrak{r}}} L\Omega.$$

PROOF. Define S as in Lemma 10.1. Define $B = A\{\mathfrak{r}\} - S$. For each $w \in \ddot{T}_{\mathfrak{r}}$ define $B_w = B \cap \mathbf{P}^{-1}(w)$. Assume (a). Then $\mathbf{P}: B \to \ddot{T}_{\mathfrak{r}}$ is a local diffeomorphism. Hence

$$\infty > \int_B \nu |g| \xi = \int_{\ddot{T}_\mathfrak{r}} \Big(\sum_{\mathfrak{z} \in B_w} \nu(\mathfrak{z}) |g(\mathfrak{z})| \Big) \ddot{\xi}(w)$$

where $L_0(w) = \Sigma_{\mathfrak{z} \in B_w} \nu(\mathfrak{z})|g(\mathfrak{z})|$ converges almost everywhere on $\ddot{T}_\mathfrak{r}$. If $w \in \ddot{T}_\mathfrak{r} - \mathbf{P}(S)$, take $\mathfrak{a} \in T_\mathfrak{r}$ with $\mathbf{P}(\mathfrak{a}) = w$. Then $B_w = \{u\mathfrak{a} \mid |u| < R\} \cap A$ and $\nu(u\mathfrak{a}) = \nu(u;\mathfrak{a})$ if $u \in \mathbf{C}(R)$. Hence

$$L_0(w) = \sum_{|u|<R} \nu(u;\mathfrak{a}) |g(u\mathfrak{a})|.$$

Therefore (b) holds. This process can be reversed; hence (b) implies (a). Therefore (a) and (b) hold now, which implies immediately that $L(\mathfrak{a})$ converges absolutely almost everywhere on $T_\mathfrak{r}$. Also ${}^\bullet L^\bullet$ exists almost everywhere on $\ddot{T}_\mathfrak{r}$ such that $L = {}^\bullet L^\bullet \circ \mathbf{P}$ almost everywhere on $T_\mathfrak{r}$. Moreover

$$\int_{A\{\mathfrak{r}\}} \nu g \xi = \int_B \nu g \xi = \int_{\ddot{T}_\mathfrak{r}} \sum_{\mathfrak{z} \in B_w} \nu(\mathfrak{z}) g(\mathfrak{z}) \ddot{\xi}$$

$$= \int_{\ddot{T}_\mathfrak{r} - \mathbf{P}(S)} {}^\bullet L^\bullet \ddot{\xi} = \int_{\ddot{T}_\mathfrak{r}} {}^\bullet L^\bullet \ddot{\xi} = \int_{T_\mathfrak{r}} L\Omega,$$

using Lemma 10.1. q. e. d.

Again, assume $0 < R \leqslant +\infty$. Take $\mathfrak{r} \in \mathbf{R}^{+m}$. Let ν be a nonnegative divisor on $P_\mathfrak{r}(\cdot R \cdot)$. Define $A = \operatorname{supp} \nu$. Assume $0 \notin A \neq \varnothing$. Define the counting function for $\mathfrak{r}$ by

$$n_{\nu\mathfrak{r}}(r) = \int_{T_\mathfrak{r}} n_\nu(r; \mathfrak{a}/|\mathfrak{a}|)\Omega(\mathfrak{a}) \geqslant 0$$

for $0 \leqslant r < R$. Define the valence function of ν for $\mathfrak{r}$ by

$$N_{\nu\mathfrak{r}}(r) = \int_{T_\mathfrak{r}} N_\nu(r; \mathfrak{a}/|\mathfrak{a}|)\Omega(\mathfrak{a}) \geqslant 0$$

for $0 \leqslant r < R$. Because $N_\nu(r;\mathfrak{a})$ is a continuous function of $(r, \mathfrak{a})$ for $|r\mathfrak{a}| < R$, the integral $N_{\nu_\mathfrak{r}}(r)$ exists and is an increasing, continuous function of r on $\mathbf{R}[0, R)$. Because

$$n_\nu(r;\mathfrak{a}) = r \frac{\partial}{\partial r} N_\nu(r;\mathfrak{a})$$

for all $|r\mathfrak{a}| < R$. The function $n_\nu(r;\mathfrak{a}/|\mathfrak{a}|)$ is measurable on $T_\mathfrak{r}$. Take $\eta > 1$ such that $r\eta^2 < R$. Then

$$\eta_\nu(r;\mathfrak{a}) \log \eta \leqslant \int_r^{\eta r} n_\nu(t;\mathfrak{a}) \frac{dt}{t} \leqslant N_\nu(\eta r;\mathfrak{a})$$

$$\leqslant (\eta + 1) \left(\frac{\eta}{\eta - 1} \right)^{2m-1} N_\nu(\eta^2 r) < \infty.$$

Hence $n_\nu(r;\mathfrak{a}/|\mathfrak{a}|)$ is bounded and measurable on $T_\mathfrak{r}$. Therefore the integral $n_{\nu_\mathfrak{r}}(r)$ exists and is an increasing function of $r \in \mathbf{R}[0, R)$.

An exchange of integration shows easily

$$N_{\nu\mathfrak{r}}(r) = \int_0^r n_{\nu\mathfrak{r}}(t) \frac{dt}{t}.$$

Now, assume $|\mathfrak{r}| = 1$ in addition. Then Lemma 10.5 implies immediately

$$n_{\nu\mathfrak{r}}(r) = \int_{A[\mathfrak{r};r]} \nu\xi, \quad N_{\nu\mathfrak{r}}(r) = \int_{A[\mathfrak{r};r]} \nu \log \frac{r}{\tau_0} \xi.$$

Also, Lemma 10.3 implies

$$n_\nu(r) = 2^m \int_0^\infty \cdots \int_0^\infty n_{\nu\mathfrak{x}}(r) e^{-|\mathfrak{x}|^2} x_1 \cdots x_m \, dx_1 \cdots dx_m,$$

$$N_\nu(r) = 2^m \int_0^\infty \cdots \int_0^\infty n_{\nu\mathfrak{x}}(r) e^{-|\mathfrak{x}|^2} x_1 \cdots x_m \, dx_1 \cdots dx_m,$$

where $\mathfrak{x} = (x_1, \cdots, x_m)$.

Let u be a pluri-subharmonic function on $P_{\mathfrak{x}}(\cdot R \cdot)$. For $\mathfrak{r} = (r_1, \cdots, r_m) \in \mathbf{R}^{+m}$ with $r_\mu < R x_\mu$ for $\mu = 1, \cdots, m$ define

$$\Phi(r_1, \cdots, r_m) = \frac{1}{(2\pi)^m} \int_0^{2\pi} \cdots \int_0^{2\pi} u(r_1 e^{i\phi_1}, \cdots, r_m e^{i\phi_m}) \, d\phi_1 \cdots d\phi_m.$$

Then Φ is an increasing function of each variable r_μ for $\mu = 1, \cdots, m$.

THEOREM 10.6 (RONKIN [35]). *Take* $0 < R \leqslant +\infty$. *Take* $\mathfrak{x} = (x_1, \cdots, x_m) \in \mathbf{R}^{+m}$. *Define* $x_0 = \mathrm{Min}(x_1, \cdots, x_m)$. *Let* $\nu \geqslant 0$ *be a nonnegative divisor on* $P_{\mathfrak{x}}(\cdot R \cdot)$. *Let* θ *and* r *be real numbers with* $0 < \theta < 1$ *and* $0 < r < \theta R$. *Then*

$$N_\nu\left(\frac{x_0}{|\mathfrak{x}|} r\right) \leqslant N_{\nu\mathfrak{x}}(r) \leqslant \frac{1+\theta}{(1-\theta)^{2m-1}} N_\nu\left(\frac{r}{\theta}\right).$$

If $R = +\infty$, *then* $\mathrm{Ord}\, \nu = \mathrm{Ord}\, N_{\nu\mathfrak{x}} = \mathrm{Ord}\, n_{\nu\mathfrak{x}}$. *If* $\mathrm{Ord}\, \nu < +\infty$, *then* class ν = class $n_{\nu\mathfrak{x}}$ = class $N_{\nu\mathfrak{x}}$ *and* type ν = type $n_{\nu\mathfrak{x}}$ = type $N_{\nu\mathfrak{x}}$.

PROOF. If $\mathfrak{a} \in \mathbf{C}^m$, if $r \in \mathbf{R}_+$ and $s \in \mathbf{R}_+$ then

$$N_\nu(rs; \mathfrak{a}) = N_\nu(r, s\mathfrak{a})$$

as far as N_ν is observed. Define $y_\mu = x_\mu |\mathfrak{x}|^{-1}$ for $\mu = 0, 1, \cdots, m$. Take $\mathfrak{t} = (t_1, \cdots, t_m) \in \mathbf{R}^{+m}$. Define $s_\mu = t_\mu |\mathfrak{t}|^{-1}$ for $\mu = 1, \cdots, m$. Proposition 3.9 implies

$$\begin{aligned} N_{\nu\mathfrak{t}}(ry_0) &= \int_{T_\mathfrak{t}} N_\nu(ry_0; \mathfrak{a}/|\mathfrak{a}|)\, \Omega(\mathfrak{a}) \\ &= \left(\frac{1}{2\pi}\right)^m \int_0^{2\pi} \cdots \int_0^{2\pi} N_\nu(ry_0; s_1 e^{i\phi_1}, \cdots, s_m e^{i\phi_m})\, d\phi_1 \cdots d\phi_m \\ &= \left(\frac{1}{2\pi}\right)^m \int_0^{2\pi} \cdots \int_0^{2\pi} N_\nu(r; y_0 s_1 e^{i\phi_1}, \cdots, y_0 s_m e^{i\phi_m})\, d\phi_1 \cdots d\phi_m \\ &\leqslant \left(\frac{1}{2\pi}\right)^m \int_0^{2\pi} \cdots \int_0^{2\pi} N_\nu(r; y_1 e^{i\phi_1}, \cdots, y_m e^{i\phi_m})\, d\phi_1 \cdots d\phi_m \\ &= \int_T N_\nu(r; \mathfrak{a}/|\mathfrak{a}|)\, \Omega(\mathfrak{a}) = N_{\nu\mathfrak{x}}(r). \end{aligned}$$

Now

$$N_\nu(ry_0) = 2^m \int_0^\infty \cdots \int_0^\infty N_{\nu\mathfrak{t}}(ry_0) e^{-|\mathfrak{t}|^2} t_1 \cdots t_m \, dt_1 \cdots dt_m \leqslant N_{\nu\mathfrak{x}}(r).$$

Also, Proposition 3.9 implies

$$N_{\nu\mathfrak{x}}(r) = \int_{T_\mathfrak{x}} N_\nu(r; \mathfrak{a}/|\mathfrak{a}|)\, \Omega(\mathfrak{a}) \leqslant \frac{1+\theta}{(1-\theta)^{2m-1}} N_\nu\left(\frac{r}{\theta}\right).$$

Now the remaining assertions follow trivially. q.e.d.

LEMMA 10.7. *Assume* $0 < s < r < R \leqslant +\infty$. *Take* $\mathfrak{x} \in \mathbf{R}^{+m}$ *with* $|\mathfrak{x}| = 1$. *Let* ν

be a divisor on $P_{\mathfrak{x}}(\cdot R \cdot)$. *Define* $A = \operatorname{supp} \nu$. *Assume* $0 \notin A$. *Let* $g : \mathbf{R}[s, r] \to \mathbf{C}$ *be a function of class* C^1. *Define* $B = A(\mathfrak{x}, r) - A(\mathfrak{x}, s)$. *Define*

$$\int_B \nu g \circ \tau_0 \xi = n_{\nu \mathfrak{x}}(r) g(r) - n_{\nu \mathfrak{x}}(s) g(s) - \int_s^r n_{\nu \mathfrak{x}}(t) g'(t)\, dt.$$

PROOF. The function g extends to a function $g : \mathbf{R} \to \mathbf{C}$ of class C^1. Then

$$\int_{A(\mathfrak{x},r)} \nu g \circ \tau_0 \xi = \int_{T_{\mathfrak{x}}} \sum_{|u|<r} \nu(u; \mathfrak{a}) g(|\mathfrak{a}|) \Omega(\mathfrak{a})$$

$$= \int_{T_{\mathfrak{x}}} \left[n_\nu(r; \mathfrak{a}) - \int_0^r n_\nu(t; \mathfrak{a}) g'(t)\, dt \right] \Omega(\mathfrak{a})$$

$$= n_{\nu \mathfrak{x}}(r) - \int_0^r n_{\nu \mathfrak{x}}(t) g'(t)\, dt$$

for all $0 < r < R$. Subtraction proves the lemma. q. e. d.

By standard methods, similar to [53, Lemma 5.6] the following lemma is derived.

LEMMA 10.8. *Assume* $0 < s < R \leqslant +\infty$. *Take* $\mathfrak{x} \in \mathbf{R}^{+m}$ *with* $|\mathfrak{x}| = 1$. *Let* $\nu \geqslant 0$ *be a nonnegative divisor on* $P_{\mathfrak{x}}(\cdot R \cdot)$. *Define* $A = \operatorname{supp} \nu$. *Assume* $0 \notin A$. *Let* g: $\mathbf{R}[s, R) \to \mathbf{R}_+$ *be a decreasing function of class* C^1 *such that* $g(r) \to 0$ *for* $r \to R$. *Define* $B = A\{\mathfrak{x}\} - A(\mathfrak{x}, s)$. *Assume either one of the two following integrals exist, then both exist and*

$$\int_B \nu g \circ \tau_0 \xi = - n_{\nu \mathfrak{x}}(s) g(s) - \int_s^R n_{\nu \mathfrak{x}}(t) g'(t)\, dt.$$

Moreover $n_{\nu \mathfrak{x}}(r) g(r) \to 0$ *as* $r \to R$.

An immediate consequence of Theorem 10.6 and Lemma 10.8 is

PROPOSITION 10.9. *Let* $\nu \geqslant 0$ *be a nonnegative divisor on* $\mathbf{C}^m$ *with* $m > 1$. *Take* $\mu > 0$. *Take* $\mathfrak{x} \in \mathbf{R}^{+m}$ *with* $|\mathfrak{x}| = 1$. *Define* $A = \operatorname{supp} \nu$. *Assume* $0 \notin A$. *Then the following statements are equivalent*:

(a) *The number* μ *is an exponent of convergence* ν.

(b) *The integral* $\int_0^\infty n_\nu(t) t^{-\mu-1}\, dt$ *exists.*

(c) *The integral* $\int_0^\infty N_\nu(t) t^{-\mu-1}\, dt$ *exists.*

(d) *The integral* $\int_0^\infty N_{\nu \mathfrak{x}}(t) t^{-\mu-1}\, dt$ *exists.*

(e) *The integral* $\int_0^\infty n_{\nu \mathfrak{x}}(t) t^{-\mu-1}\, dt$ *exists.*

(f) *The integral* $\int_{A\{\mathfrak{x}\}} \nu \tau_0^{-\mu} \xi$ *exists.*

Now, the Jensen formula, the First Main Theorem and the Jensen-Poisson formula shall be derived.

THEOREM 10.10. JENSEN'S FORMULA. *Take* $0 < r < R \leqslant +\infty$. *Take* $\mathfrak{x} \in \mathbf{R}^{+m}$ *with* $|\mathfrak{x}| = 1$. *Let* f *be a meromorphic function on* $P_{\mathfrak{x}}(\cdot R \cdot)$ *which is holomorphic at* $0 \in \mathbf{C}^m$ *and such that* $f(0) \neq 0$. *Then*

$$\int_{T_{\mathfrak{x}}} \log |f(r\mathfrak{y})| \,\Omega(\mathfrak{y}) = N_{\nu \mathfrak{x}}(r) + \log |f(0)|.$$

PROOF.

$$N_{f\mathfrak{x}}(r) + \log |f(0)| = \int_{T_\mathfrak{x}} (N_f(r;\mathfrak{a}) + \log |f(0)|)\Omega(\mathfrak{a})$$

$$= \int_{T_\mathfrak{x}} \frac{1}{2\pi} \int_0^{2\pi} \log |f(re^{i\phi}\mathfrak{a})|\, d\phi\,\Omega(\mathfrak{a})$$

$$= \int_{T_\mathfrak{x}} \log |f(r\mathfrak{y})|\,\Omega(\mathfrak{y}). \quad \text{q. e. d.}$$

Let V be a hermitian vector space of dimension $n + 1 > 1$. Take $\mathfrak{x} \in \mathbf{R}^{+m}$. Let

$$f : P_\mathfrak{x}(\cdot R \cdot) \longrightarrow \mathbf{P}(V)$$

be a meromorphic map. Assume that f is holomorphic at $0 \in \mathbf{C}^m$. The characteristic of f for $\mathfrak{x}$ is defined by

$$T_{f\mathfrak{x}}(r) = \int_{T_\mathfrak{x}} T_f(r; \mathfrak{a}/|\mathfrak{a}|)\Omega(\mathfrak{a}).$$

Then $T_{f\mathfrak{x}}$ is an increasing, continuous function on $\mathbf{R}[0, R)$. Again, Lemma 10.3 implies

$$T_f(r) = 2^m \int_0^\infty \cdots \int_0^\infty T_{f\mathfrak{x}}(r) e^{-|\mathfrak{x}|^2} x_1 \cdots x_m \, dx_1 \cdots dx_m.$$

Define $x_0 = \operatorname{Min}(x_1, \cdots, x_m)$. Take $0 < \theta < 1$ and $0 < r < \theta R$. The same reasoning as in the proof of Theorem 10.6 implies

$$T_f\left(\frac{x_0}{|\mathfrak{x}|} r\right) \leqslant T_{f\mathfrak{x}}(r) \leqslant \frac{1+\theta}{(1-\theta)^{2m-1}} T_f\left(\frac{r}{\theta}\right).$$

Hence, if $R = \infty$, then $\operatorname{Ord} f = \operatorname{Ord} T_f$. If $\operatorname{Ord} f < \infty$, then class f = class T_f and type f = type T_f.

Take $a \in \mathbf{P}(V^*)$. Assume that $f(0) \notin \ddot{E}[a]$. Define

$$m^a_{f\mathfrak{x}}(r) = \int_{T_\mathfrak{x}} m^a_f(r; \mathfrak{a}/|\mathfrak{a}|)\,\sigma(\mathfrak{a}) \quad \text{if } 0 < r < R,$$

$$m^a_{f\mathfrak{x}}(0) = m^a_f(0) = \log(1/\|f(0), a\|).$$

Obviously, the First Main Theorem follows

$$T_{f\mathfrak{x}}(r) = N^a_{f\mathfrak{x}}(r) + m^a_{f\mathfrak{x}}(r) - m^a_{f\mathfrak{x}}(0).$$

If $|\mathfrak{x}| = 1$, then Lemma 10.2 implies

$$m^a_{f\mathfrak{x}}(r) = \int_{T_\mathfrak{x}\langle\cdot r \cdot\rangle} \log(1/\|f, a\|)\Omega.$$

The primary concern is the construction of canonical functions. Hence this line of investigation of meromorphic maps shall not be continued. The fundamental Jensen-Poisson formula for polydiscs shall be derived. This necessitates quite a number of notations.

For $\mathfrak{z} = (z_1, \cdots, z_m)$ and $\mathfrak{w} = (w_1, \cdots, w_m)$ in $\mathbf{C}^m$ define

$$(\mathfrak{z}|\mathfrak{w}) = \sum_{\mu=1}^m z_\mu \bar{w}_\mu,$$

$$|\mathfrak{z}| = \left(\sum_{\mu=1}^m |z_\mu|^2\right)^{1/2}, \qquad \|\mathfrak{z}\| = \operatorname{Max}(|z_1|, \cdots, |z_m|),$$

$$\mathfrak{z}\mathfrak{w} = (z_1 w_1, \cdots, z_m w_m), \quad \mathfrak{z}/\mathfrak{w} = (z_1/w_1, \cdots, z_m/w_m) \text{ if } \mathfrak{w} \in \mathbf{C}_*^m,$$

$$\mathfrak{e} = (1, \cdots, 1) \in \mathbf{C}^m, \quad \mathfrak{z}\mathfrak{e} = \mathfrak{e}\mathfrak{z} = \mathfrak{z},$$

$$\uparrow \mathfrak{z} = (\mathfrak{z} \mid \mathfrak{e}) = z_1 + \cdots + z_m.$$

If $\mathfrak{p} = (p_1, \cdots, p_m) \in \mathbf{Z}^m$, define

$$\mathfrak{z}^{\mathfrak{p}} = z_1^{p_1} \cdots z_m^{p_m}$$

where $z_\mu^{p_\mu} = 1$ if $p_\mu = 0$ and $z_\mu \in \mathbf{C}$. If $p_\mu < 0$, assume $z_\mu \neq 0$. For instance

$$\mathfrak{z}^{\mathfrak{e}} = z_1 \cdots z_m, \quad (\mathfrak{e} - \mathfrak{z})^{-\mathfrak{e}} = \prod_{\mu=1}^{m} \frac{1}{1 - z_\mu}.$$

For $p \in \mathbf{Z}_+$, define

$$\mathbf{D}_p = \{\mathfrak{p} \in \mathbf{Z}_+^m \mid \uparrow \mathfrak{p} = p\}.$$

A homogeneous polynomial Q_p of degree p on $\mathbf{C}^m$ is defined by

$$Q_p(\mathfrak{z}) = \sum_{\mathfrak{p} \in \mathbf{D}_p} \mathfrak{z}^{\mathfrak{p}}.$$

Then

$$(\mathfrak{e} - \mathfrak{z})^{-\mathfrak{e}} = \sum_{p=0}^{\infty} Q_p(\mathfrak{z})$$

for all $\mathfrak{z} \in T_{\mathfrak{e}}$. The infinite series converges absolutely and uniformly on each compact subset of $T_{\mathfrak{e}}$. Observe

$$Q_p(\mathfrak{e}) = \# \mathbf{D}_p = \binom{m+p}{p}, \quad |Q_p(\mathfrak{z})| \leq \binom{m+p}{p} \|\mathfrak{z}\|^p.$$

For each $q \in \mathbf{Z}_+$ define a holomorphic function $\Lambda(\Box, q)$ on $T_{\mathfrak{e}}$ by

$$\Lambda(\mathfrak{z}, q) = \int_0^1 \left[\sum_{p=0}^{q} Q_p(\mathfrak{z}) t^p - (\mathfrak{e} - t\mathfrak{z})^{-\mathfrak{e}} \right] \frac{dt}{t},$$

$$\Lambda(\mathfrak{z}) = \Lambda(\mathfrak{z}, 0) = \int_0^1 [1 - (\mathfrak{e} - t\mathfrak{z})^{-\mathfrak{e}}] \frac{dt}{t}.$$

Observe that $1/t$ cancels. Hence the integrals exist. Then

$$\Lambda(\mathfrak{z}, q) = -\sum_{p=q+1}^{\infty} \frac{1}{p} Q_p(\mathfrak{z})$$

for $\mathfrak{z} \in T_{\mathfrak{e}}$. If $H \neq \emptyset$ is an open subset of $\mathbf{C}^m$ and if $f: H \to \mathbf{C}$ is holomorphic, define

$$h'(\mathfrak{z}) = \frac{d}{dx} f(x\mathfrak{z}) \Big|_{x=1} = \sum_{\mu=1}^{m} z_\mu f_{z_\mu}(\mathfrak{z}).$$

Then

$$\Lambda'(\mathfrak{z}, q) = \sum_{p=q+1}^{\infty} Q_p(\mathfrak{z}) = \sum_{p=0}^{q} Q_p(\mathfrak{z}) - (\mathfrak{e} - \mathfrak{z})^{-\mathfrak{e}}.$$

LEMMA 10.11. *Take* $q \in \mathbf{Z}_+$ *and* $\mathfrak{z} \in T_{\mathfrak{e}}$. *Then*

$$|\Lambda(\mathfrak{z}, q)| \leq (q+2)^m \|\mathfrak{z}\|^{q+1} (1 - \|\mathfrak{z}\|)^{-m},$$

$$|\Lambda'(\mathfrak{z}, q)| \leq \frac{(q+2)^m}{q+1} \|\mathfrak{z}\|^{q+1} (1 - \|\mathfrak{z}\|)^{-m}.$$

PROOF.

$$|\Lambda'(\mathfrak{z}, q)| \leqslant \sum_{p=q+1}^{\infty} \binom{p+m}{p} \|\mathfrak{z}\|^p = \sum_{p=0}^{\infty} \binom{p+m+q+1}{p} \|\mathfrak{z}\|^{p+q+1}$$

$$\leqslant (q+2)^m \|\mathfrak{z}\|^{q+1} \sum_{p=0}^{\infty} \binom{p+m}{p} \|\mathfrak{z}\|^p$$

$$= (q+2)^m \|\mathfrak{z}\|^{q+1} (1 - \|\mathfrak{z}\|)^{-m};$$

$$|\Lambda(\mathfrak{z}, q)| \leqslant \sum_{p=q+1}^{\infty} \binom{p+m}{p} \frac{1}{p} \|\mathfrak{z}\|^p \leqslant \frac{1}{q+1} \sum_{p=q+1}^{\infty} \binom{p+m}{p} \|\mathfrak{z}\|^p$$

$$\leqslant \frac{(q+2)^m}{q+1} \|\mathfrak{z}\|^{q+1} (1 - \|\mathfrak{z}\|)^{-m}. \quad \text{q. e. d.}$$

LEMMA 10.12. *Take* $\mathfrak{x} \in \mathbf{R}^{+m}$. *Let* F *be a homogeneous polynomial of degree* n *on* $\mathbf{C}^m$. *Take* $p \in \mathbf{Z}_+$. *Then*

$$\int_{T_{\mathfrak{x}}} F(\mathfrak{y}) Q_p(\mathfrak{z}/\mathfrak{y}) \,\Omega(\mathfrak{y}) = \begin{cases} F(\mathfrak{z}) & \text{if } p = n, \\ 0 & \text{if } p \neq n. \end{cases}$$

PROOF. For each $\mathfrak{n} \in \mathbf{D}_n$ a complex number $a_{\mathfrak{n}}$ exists such that

$$F(\mathfrak{y}) = \sum_{\mathfrak{n} \in \mathbf{D}_n} a_{\mathfrak{n}} \mathfrak{y}^{\mathfrak{n}} \quad \text{for all } \mathfrak{y} \in \mathbf{C}^m.$$

Then

$$F(\mathfrak{y})\, Q(\mathfrak{z}/\mathfrak{y}) = \sum_{\mathfrak{n} \in \mathbf{D}_n} \sum_{\mathfrak{p} \in \mathbf{D}_p} a_{\mathfrak{n}} \mathfrak{y}^{\mathfrak{n}-\mathfrak{p}} \mathfrak{z}^{\mathfrak{n}}$$

for all $\mathfrak{y} \in \mathbf{C}^m_*$ and $\mathfrak{z} \in \mathbf{C}^m$. Then

$$\int_{T_{\mathfrak{x}}} \mathfrak{y}^{\mathfrak{n}-\mathfrak{p}} \,\Omega(\mathfrak{y}) = \left(\frac{1}{2\pi}\right)^m \int_0^{2\pi} \cdots \int_0^{2\pi} e^{\uparrow(\mathfrak{n}-\mathfrak{p})\cdot\phi i} \mathfrak{x}^{\mathfrak{n}-\mathfrak{p}} \, d\phi_1 \cdots d\phi_m$$

$$= \begin{cases} 1 & \text{if } \mathfrak{n} = \mathfrak{p}, \\ 0 & \text{if } \mathfrak{n} \neq \mathfrak{p}, \end{cases}$$

which immediately implies the assertion of the lemma. q. e. d.

An immediate consequence is the well-known Cauchy integral formula: Assume $0 < r < R \leqslant +\infty$. Let $f : P_{\mathfrak{x}}(\cdot R \cdot) \to \mathbf{C}$ be a holomorphic function. Take $\mathfrak{z} \in P_{\mathfrak{x}}(\cdot r \cdot)$. Then

$$F(\mathfrak{z}) = \int_{T_{\mathfrak{x}}} F(r\mathfrak{y}) \, [1 - \Lambda'(\mathfrak{z}/(r\mathfrak{y}), q)] \,\Omega(\mathfrak{y}),$$

THEOREM 10.13. JENSEN-POISSON FORMULA FOR POLYDISCS. *Take* $0 < R \leqslant +\infty$. *Take* $\mathfrak{x} \in \mathbf{R}^{+m}$ *with* $|\mathfrak{x}| = 1$. *Let* f *be a meromorphic function on* $P_{\mathfrak{x}}(\cdot R \cdot)$. *Assume that* f *is holomorphic at* $0 \in \mathbf{C}^m$ *with* $f(0) = 1$. *Define* $\nu = \mu_f$ *and* $A = \operatorname{supp} \mu_f$. *Assume* $A \neq \varnothing$. *Let* s *be maximal with* $0 < s < R$ *such that* $A \cap P_{\mathfrak{x}}(\cdot s \cdot) = \varnothing$. *Then* $\log f = \sum_{p=1}^{\infty} F_p$ *on* $P_{\mathfrak{x}}(\cdot s \cdot)$ *where* F_p *is a homogeneous polynomial of degree* p. *Take* $q \in \mathbf{Z}_+$. *Take* $r \in \mathbf{R}[s, R)$. *Then holomorphic functions* F *and* E *on* $P_{\mathfrak{x}}(\cdot r \cdot)$ *are defined by*

$$F(\mathfrak{z}) = \int_{T_{\mathfrak{x}}} \log |f(r\mathfrak{y})| \,\Lambda'(\mathfrak{z}/(r\mathfrak{y}), q) \,\Omega(\mathfrak{y}),$$

$$E(\mathfrak{z}) = \int_{A(\mathfrak{r},r)} \nu(\mathfrak{y})\Lambda(\mathfrak{z}\bar{\mathfrak{y}}/(r^2\mathfrak{r}^2), q)\xi(\mathfrak{y}),$$

for all $\mathfrak{z} \in P_{\mathfrak{r}}(\cdot r \cdot)$. *Also, a holomorphic function* H *on* $P_{\mathfrak{r}}(\cdot s \cdot)$ *is defined by*

$$H(\mathfrak{z}) = \int_{A(\mathfrak{r},r)} \nu(\mathfrak{y})\Lambda(\mathfrak{z}/\mathfrak{y}, q)\xi(\mathfrak{y})$$

for all $\mathfrak{z} \in P_{\mathfrak{r}}(\cdot s \cdot)$. *Moreover*

$$\log f = \sum_{p=1}^{q} F_p - 2F + H - E.$$

PROOF. Take $0 < \theta < 1$. If $\mathfrak{z} \in P_{\mathfrak{r}}(\theta r)$ and $\mathfrak{y} \in T_{\mathfrak{r}}$, then $\|\mathfrak{z}/r\mathfrak{y}\| \leqslant \theta < 1$ and $\|\mathfrak{z}\bar{\mathfrak{y}}/r^2\mathfrak{r}^2\| \leqslant \theta < 1$. Hence

$$|\Lambda'(\mathfrak{z}/r\mathfrak{y}, q)| \leqslant (q+2)^m \frac{\theta^{q+1}}{(1-\theta)^m}, \qquad |\Lambda(\mathfrak{z}\bar{\mathfrak{y}}/r^2\mathfrak{r}^2, q)| \leqslant \frac{(q+2)^m}{(q+1)} \frac{\theta^{q+1}}{(1-\theta)^m}.$$

Therefore F and E are holomorphic on $P_{\mathfrak{r}}(\cdot r \cdot)$.

Take $0 < \theta < 1$. Take $\mathfrak{z} \in P_{\mathfrak{r}}(\cdot \theta s \cdot)$ and $\mathfrak{y} \in A(\mathfrak{r}, r)$. Then $\|\mathfrak{z}/\mathfrak{y}\| \leqslant \theta < 1$. Therefore

$$|\Lambda(\mathfrak{z}/\mathfrak{y}, q)| \leqslant \frac{(q+2)^m}{q+1} \frac{\theta^{q+1}}{(1-\theta)^m}.$$

Therefore H is holomorphic on $P_{\mathfrak{r}}(\cdot s \cdot)$.

Take $\mathfrak{a} \in T_{\mathfrak{r}}$ and $z \in \mathbf{C}(s)$. Then (5.1) implies

$$\begin{aligned}\log f(z\mathfrak{a}) = {} & 2\sum_{p=1}^{\infty} \frac{1}{2\pi}\int_0^{2\pi} \log|f(re^{i\phi}\mathfrak{a})|\,e^{-ip\phi}\,d\phi (z/r)^p \\ & - \sum_{p=1}^{\infty} \frac{1}{p}\Big(\sum_{|u|<r} \nu(u;\mathfrak{a})u^{-p}\Big) z^p \\ & + \sum_{p=1}^{\infty} \frac{1}{p}\Big(\sum_{|u|<r} \nu(u;\mathfrak{a})\bar{u}^p\Big)(z/r^2)^p.\end{aligned}$$

Therefore

$$\begin{aligned}F_p(\mathfrak{a}) = {} & \frac{1}{\pi} r^{-p}\int_0^{2\pi} \log|f(re^{i\phi}\mathfrak{a})|\,e^{-ip\phi}\,d\phi \\ & - \frac{1}{p}\sum_{|u|<r} \nu(u;\mathfrak{a})u^{-p} + \frac{1}{p}\sum_{|u|<r} \nu(u;\mathfrak{a})(\bar{u}/r^2)^p.\end{aligned} \tag{10.1}$$

Take $\mathfrak{z} \in \mathbf{C}^m$. Then

$$\begin{aligned}F_p(\mathfrak{z}) &= \int_{T_{\mathfrak{r}}} F_p(\mathfrak{a})Q_p(\mathfrak{z}/\mathfrak{a})\Lambda(\mathfrak{a}) \\ &= \frac{1}{2\pi}\int_0^{2\pi} 2\int_{T_{\mathfrak{r}}} \log|f(re^{i\phi}\mathfrak{a})|\,Q_p(\mathfrak{z}/r\mathfrak{a})\,e^{-ip\phi}\,\Lambda(\mathfrak{a})\,d\phi \\ &\quad - \frac{1}{p}\int_{T_{\mathfrak{r}}} \sum_{|u|<r} \nu(u;\mathfrak{a})Q_p(\mathfrak{z}/u\mathfrak{a})\Lambda(\mathfrak{a}) \\ &\quad + \frac{1}{p}\int_{T_{\mathfrak{r}}} \sum_{|u|<r} \nu(u;\mathfrak{a})Q_p(\mathfrak{z}\bar{u}\bar{\mathfrak{a}}/r^2\mathfrak{r}^2)\Lambda(\mathfrak{a})\end{aligned}$$

$$= 2\int_{T_{\mathfrak{x}}} \log |f(r\mathfrak{y})| Q_p(\mathfrak{z}/(r\mathfrak{z}))\Lambda(\mathfrak{y})$$

$$-\frac{1}{p}\int_{A(\mathfrak{x},r)} \nu(\mathfrak{y}) Q_p(\mathfrak{z}/\mathfrak{y})\,\xi(\mathfrak{y})$$

$$+\frac{1}{p}\int_{A(\mathfrak{x},r)} \nu(\mathfrak{y}) Q_p(\mathfrak{z}\bar{\mathfrak{y}}/(r^2\mathfrak{x}^2))\xi(\mathfrak{y}).$$

Summation over all $p \in \mathbf{N}[q+1, +\infty)$ implies the Jensen-Poisson formula. q. e. d.

If ν is a divisor on $P_{\mathfrak{x}}(\cdot R \cdot)$ with $A = \operatorname{supp} \nu$ and $0 \notin A \neq \varnothing$, then a meromorphic function f on $P_{\mathfrak{x}}(\cdot R \cdot)$ exists such that $\mu_f = \nu$ and $f(0) = 1$. Then

$$h = f \exp\left(-\sum_{p=1}^{q} F_p + 2F + E\right)$$

is a meromorphic function on $P_{\mathfrak{x}}(\cdot r \cdot)$ with $\mu_h = \nu$. Also h is holomorphic on $P_{\mathfrak{x}}(\cdot s \cdot)$ with $h(0) = 1$ such that $\log h = H$. Hence h does not depend on the choice of f. This implies the following result.

THEOREM 10.14. *Take* $0 < R \leqslant +\infty$. *Take* $\mathfrak{x} \in \mathbf{R}^{+m}$ *with* $|\mathfrak{x}| = 1$. *Let* ν *be a divisor on* $P_{\mathfrak{x}}(\cdot R \cdot)$. *Define* $A = \operatorname{supp} \nu$. *Assume* $0 \notin A \neq \varnothing$. *Let* s *be maximal with* $A \cap P_{\mathfrak{x}}(\cdot s \cdot) = \varnothing$. *Take* $q \in \mathbf{Z}_+$. *Take* $r \in \mathbf{R}[s, R)$. *Then there exists one and only one meromorphic function* h *on* $P_{\mathfrak{x}}(\cdot r \cdot)$ *such that* h *is holomorphic on* $P_{\mathfrak{x}}(\cdot s \cdot)$ *with* $h(0) = 1$ *and such that*

$$\log h(\mathfrak{z}) = \int_{A(\mathfrak{x},r)} \nu(\mathfrak{y})\Lambda(\mathfrak{z}/\mathfrak{y}, q)\,\xi(\mathfrak{y})$$

for all $\mathfrak{z} \in P_{\mathfrak{x}}(s)$. *Moreover,* $\mu_h = \nu | P_{\mathfrak{x}}(\cdot r \cdot)$ *is the divisor of* h.

Now, the Weierstrass integral theorem shall be derived for a polydisc exhaustion.

Let $u : \mathbf{R}_+ \to \mathbf{R}_+$ be an increasing function. Then an increasing function $B : \mathbf{R}_+ \to \mathbf{R}_+$ of class C^∞ is said to be a ballast function if and only if for every $r \in \mathbf{R}^+$ the following integral exists:

$$\int_r^\infty u(t)\,(r/t)^{B(t)+2}\,(B(t) + 1 + B'(t)\,t \log(t/r))\,dt.$$

In [53, Lemma 5.7 and Lemma 5.8], the existence of an unbounded ballast function B with $B \geqslant u$ was shown.

Take $\mathfrak{x} \in \mathbf{R}^{+m}$ with $|\mathfrak{x}| = 1$. Let $\nu \geqslant 0$ be a nonnegative divisor on $\mathbf{C}^m$. Define $A = \operatorname{supp} \nu$. Assume $0 \notin A \neq \varnothing$. Then an increasing function $q : \mathbf{R}_+ \to \mathbf{Z}_+$ is said to be a weight function for ν and $\mathfrak{x}$ if and only if the integral

$$\int_{A\{\mathfrak{x}\}} \nu (r/\tau_0)^{q\circ\tau_0+1}\,\xi < \infty$$

exists for each $r > 0$.

LEMMA 10.15. *Take* $\mathfrak{x} \in \mathbf{R}^{+m}$ *with* $|\mathfrak{x}| = 1$. *Let* $\nu \geqslant 0$ *be a nonnegative divisor on* $\mathbf{C}^m$. *Define* $A = \operatorname{supp} \nu$. *Assume* $0 \notin A \neq \varnothing$. *Let* q *be a ballast function for* $n_{\nu\mathfrak{x}}$. *Then* $q_0 : \mathbf{R}_+ \to \mathbf{Z}_+$ *is uniquely defined by*

$$q_0(r) - 1 < q(r) \leqslant q_0(r), \quad \textit{for all } r \geqslant 0,$$

and q_0 *is a weight function of* ν *for* $\mathfrak{x}$. *Especially, a weight function of* ν *for* $\mathfrak{x}$ *exists.*

PROOF. Take $r > 0$. Define

$$f(t) = (r/t)^{q(t)+1}$$

for $t \geqslant r$. Then

$$f'(t) = (q'(t) \log (r/t) - (q(t) + 1)/t) f(t) \leqslant 0$$

for $t \geqslant r$. Hence f decreases. Obviously $f(t) \to 0$ for $t \to \infty$. Define $B = A\{\mathfrak{x}\} - A(\mathfrak{x}, r)$. Lemma 10.8 implies

$$\infty > \int_r^\infty n_{\nu\mathfrak{x}}(t)\, (r/t)^{q(t)+2}\, [q(t) + 1 + tq'(t) \log (t/r)]\, dt$$
$$= n_{\nu\mathfrak{x}}(r)\, (q(r) + 1) + \int_B \nu (r/\tau_0)^{q\circ\tau_0 + 1}\, \xi.$$

If $t \geqslant r$, then

$$(r/t)^{q_0(t)+1} \leqslant (r/t)^{q(t)+1}.$$

Hence

$$\int_{A\{\mathfrak{x}\}} \nu (r/\tau_0)^{q_0 \circ \tau + 1}\, \xi < \infty$$

exists. q. e. d.

LEMMA 10.16. *Take* $\mathfrak{x} \in \mathbf{R}^{+m}$ *with* $|\mathfrak{x}| = 1$. *Let* ν *be a nonnegative divisor on* $\mathbf{C}^m$. *Define* $A = \operatorname{supp} \nu$. *Assume* $0 \notin A \neq \varnothing$. *Let* s *be a maximal such that* $A \cap P_{\mathfrak{x}}(\cdot s \cdot) = \varnothing$. *Let* $q : \mathbf{R}_+ \to \mathbf{Z}_+$ *be a weight function of* ν *for* $\mathfrak{x}$. *Take* $r \geqslant s$. *Define* $B = A\{\mathfrak{x}\} - A(\mathfrak{x}, r)$. *Then the integral*

$$H(\mathfrak{z}) = \int_B \nu(\mathfrak{y})\, \Lambda(\mathfrak{z}/\mathfrak{y}, q(|\mathfrak{y}|))\, \xi(\mathfrak{y})$$

exists for each $\mathfrak{z} \in P_{\mathfrak{x}}(\cdot r \cdot)$. *The function* H *is holomorphic on* $P_{\mathfrak{x}}(\cdot r \cdot)$.

PROOF. Take any $x > 0$. The integral

$$\int_{A\{\mathfrak{x}\}} \nu (x/\tau_0)^{q\circ\tau_0 + 1}\, \xi < \infty \tag{10.2}$$

exists. Now, it is claimed that the integral

$$\int_{A\{\mathfrak{x}\}} \nu (q \circ \tau_0 + 2)^m\, (x/\tau_0)^{q\circ\tau_0 + 1}\, \xi \tag{10.3}$$

exists for all $x > 0$. If q is bounded, the existence of (10.3) follows trivially from the existence of (10.2). Assume that q is not bounded. Then

$$\frac{\log (q(t) + 2)}{q(t) + 1} \longrightarrow 0 \quad \text{for } t \longrightarrow \infty.$$

Observe that $m > 1$. Therefore a number $t_0 > s$ exists such that

$$\frac{\log m}{m} > \frac{\log (q(t) + 2)}{q(t) + 1} > 0$$

for all $t \geqslant t_0$. Hence

$$(q(t) + 2)^m\, (x/t)^{q(t)+1} \leqslant (mx/t)^{q(t)+1}$$

for all $t \geqslant t_0$. Therefore (10.3) exists.

Take $0 < \theta < 1$ and $\mathfrak{z} \in P_{\mathfrak{x}}(\cdot\,\theta r\,\cdot)$ and $\mathfrak{y} \in B$. Then

$$\|\mathfrak{z}/\mathfrak{y}\| < \theta r/|\mathfrak{y}| < \theta.$$

Hence

$$\Lambda(\mathfrak{z}/\mathfrak{y},\, q(|\mathfrak{y}|)) \leqslant q(|\mathfrak{y}| + 2)^m \frac{1}{(1-\theta)^m}\left(\frac{\theta r}{|\mathfrak{y}|}\right)^{q(|\mathfrak{y}|)+1}.$$

Therefore $H(\mathfrak{z})$ exists for each $\mathfrak{z} \in P_{\mathfrak{x}}(\cdot\,\theta r\,\cdot)$. The integral converges uniformly on $P_{\mathfrak{x}}(\cdot\,\theta r\,\cdot)$. The function H is holomorphic on $P_{\mathfrak{x}}(\cdot\,r\,\cdot)$. q. e. d.

Now, the analogue to Theorem 5.5 can be proved.

THEOREM 10.17. WEIERSTRASS INTEGRAL THEOREM. *Take $\mathfrak{x} \in \mathbf{R}^{+m}$ with $|\mathfrak{x}| = 1$ and $m > 1$. Let $\nu \geqslant 0$ be a nonnegative divisor on $\mathbf{C}^m$. Define $A = \operatorname{supp} \nu$. Assume $0 \notin A \neq \varnothing$. Take s maximal such that $A \cap P_{\mathfrak{x}}(\cdot\,s\,\cdot) = \varnothing$. Let $q: \mathbf{R}_+ \to \mathbf{Z}_+$ be a weight function of ν for $\mathfrak{x}$.*

Then there exists one and only one holomorphic function h on $\mathbf{C}^m$ such that $h(\mathfrak{z}) \neq 0$ for all $\mathfrak{z} \in P_{\mathfrak{x}}(\cdot\,r\,\cdot)$, such that $h(0) = 1$ and such that

$$\log h(\mathfrak{z}) = \int_{A\{\mathfrak{x}\}} \nu(\mathfrak{y})\Lambda(\mathfrak{z}/\mathfrak{y},\, q(|\mathfrak{y}|))\, \xi(\mathfrak{y}) \tag{10.4}$$

for all $\mathfrak{z} \in P_{\mathfrak{x}}(\cdot\,s\,\cdot)$. Moreover $\mu_h = \nu$ is the divisor of h.

PROOF. Take $r \geqslant s$. For $t \in \mathbf{R}_+$ and $p \in \mathbf{N}$ define

$$\chi_p(t, r) = \begin{cases} 1 & \text{if } p \leqslant q(t), \\ 0 & \text{if } p > q(t). \end{cases}$$

A polynomial G_r on $\mathbf{C}^m$ is defined by

$$G_r(\mathfrak{z}) = \sum_{p=1}^{q(r)} \frac{1}{p} \int_{A(\mathfrak{x},r)} \nu(\mathfrak{y})\, \chi_p(|\mathfrak{y}|, r)\, Q_p(\mathfrak{z}/\mathfrak{y})\, \xi(\mathfrak{y})$$

for all $\mathfrak{z} \in \mathbf{C}^m$. If $\mathfrak{z} \in P_{\mathfrak{x}}(\cdot\,s\,\cdot)$ and $\mathfrak{y} \in A(\mathfrak{x}, r)$, then $\|\mathfrak{z}/\mathfrak{y}\| < 1$ and

$$\sum_{p=1}^{q(r)} \frac{1}{p} \chi_p(|\mathfrak{y}|, r)\, Q_p(\mathfrak{z}/\mathfrak{y}) = \sum_{p=1}^{q(|\mathfrak{y}|)} \frac{1}{p} Q_p(\mathfrak{z}/\mathfrak{y})$$
$$= \Lambda(\mathfrak{z}/\mathfrak{y},\, q(|\mathfrak{y}|)) - \Lambda(\mathfrak{z}/\mathfrak{y}).$$

Therefore

$$G_r(\mathfrak{z}) = \int_{A(\mathfrak{x},r)} \nu(\mathfrak{y})\, \Lambda(\mathfrak{z}/\mathfrak{y},\, q(|\mathfrak{y}|))\, \xi(\mathfrak{y}) - \int_{A(\mathfrak{x},r)} \nu(\mathfrak{y})\, \Lambda(\mathfrak{z}/\mathfrak{y})\, \xi(\mathfrak{y})$$

for all $\mathfrak{z} \in P_{\mathfrak{x}}(\cdot\,s\,\cdot)$.

A holomorphic function H_r on $P_{\mathfrak{x}}(\cdot\,r\,\cdot)$ exists such that $H_r(\mathfrak{z}) \neq 0$ for all $\mathfrak{z} \in P_{\mathfrak{x}}(\cdot\,s\,\cdot)$, such that $H_r(0) = 1$ and such that

$$\log H_r(\mathfrak{z}) = \int_{A(\mathfrak{x},r)} \nu(\mathfrak{y})\, \Lambda(\mathfrak{z}/\mathfrak{y})\, \xi(\mathfrak{y})$$

for all $\mathfrak{z} \in P_{\mathfrak{x}}(\cdot\,s\,\cdot)$. Define $B = A\{\mathfrak{x}\} - A(\mathfrak{x}, r)$. A holomorphic function D_r on $P_{\mathfrak{x}}(\cdot\,r\,\cdot)$ is defined by

$$D_r(\mathfrak{z}) = \int_B \nu(\mathfrak{y})\, \Lambda(\mathfrak{z}/\mathfrak{y},\, q(|\mathfrak{y}|))\, \xi(\mathfrak{y}).$$

Then

$$h_r = H_r \exp(D_r + G_r)$$

is a holomorphic function on $P(\cdot r \cdot)$ with $\mu_{h_r} = \mu_{H_r} = \nu | P_{\mathfrak{x}}(\cdot r \cdot)$. Also $h_r(0) = 1$ and $h_r(\mathfrak{z}) \neq 0$ for all $\mathfrak{z} \in P_{\mathfrak{x}}(\cdot s \cdot)$. Moreover, addition implies

$$\log h_r(\mathfrak{z}) = \int_{A\{\mathfrak{x}\}} \nu(\mathfrak{y}) \Lambda(\mathfrak{z}/\mathfrak{y}, q(|\mathfrak{y}|)) \xi(\mathfrak{y})$$

for all $\mathfrak{z} \in P_{\mathfrak{x}}(\cdot s \cdot)$. Therefore h_r does not depend on r on $P_{\mathfrak{x}}(\cdot s \cdot)$. One and only one holomorphic function h on $\mathbf{C}^m$ exists such that $h | P_{\mathfrak{x}}(\cdot r \cdot) = h_r$ for each $r > s$. Then $h(0) = 1$ and $h(\mathfrak{z}) \neq 0$ for all $\mathfrak{z} \in P_{\mathfrak{x}}(\cdot s \cdot)$. Moreover, (10.4) holds on $P_{\mathfrak{x}}(\cdot s \cdot)$. Obviously, h is uniquely defined by these properties. Also

$$\mu_h | P_{\mathfrak{x}}(\cdot r \cdot) = \mu_{h_r} = \nu | P_{\mathfrak{x}}(\cdot r \cdot)$$

for each $r > s$. Hence $\mu_h = \nu$ on $\mathbf{C}^m$. q. e. d.

The function h is called the canonical function of ν for $\mathfrak{x}$ and for the weight function q. If $q \in \mathbf{Z}_+$ is a constant integer, such that $q + 1$ is an exponent of convergence of ν, then q is a constant weight function of ν for $\mathfrak{x}$. Now there exist, first the canonical function h of ν for the weight q as constructed in Theorem 5.5 and Theorem 6.3, and, second there exists the canonical function $h_{\mathfrak{x}}$ of ν for $\mathfrak{x}$ and for the constant weight function q. Now, it will be shown that both functions are the same: $h_{\mathfrak{x}} = h$. However the uniqueness Proposition 6.4 cannot be used for the proof, since no method is available at present to estimate $h_{\mathfrak{x}}$ directly.

LEMMA 10.18. *Let $\nu \geqslant 0$ be a nonnegative divisor on $\mathbf{C}^m$ with $0 \notin \operatorname{supp} \nu \not\equiv \varnothing$. Let $q + 1 \in \mathbf{N}$ be an exponent of convergence of ν. Let h be the canonical function of ν for the weight q. Take $0 < r < R$ and $\mathfrak{z} \in P_{\mathfrak{x}}(\cdot r \cdot)$. Then*

$$\int_{T_{\mathfrak{x}}} \log |h(R\mathfrak{y})| \Lambda'(\mathfrak{z}/(R\mathfrak{y}), q)\Omega(\mathfrak{y}) \longrightarrow \infty$$

for $R \longrightarrow \infty$ with $r < R < +\infty$. The convergence is uniform on $P_{\mathfrak{x}}(\cdot r \cdot)$.

PROOF. Define $A = \operatorname{supp} \nu$. Take s maximal with $A \cap P_{\mathfrak{x}}(\cdot s \cdot) = \varnothing$. Then $\log h = \sum_{p=q+1}^{\infty} H_p$ on $P_{\mathfrak{x}}(\cdot s \cdot)$ where H_p is a homogeneous polynomial of degree p. Take $\mathfrak{a} \in T_{\mathfrak{x}}$ and $+\infty > R > 2r$. Then (10.1) implies

$$H_p(\mathfrak{a}) = \frac{1}{\pi} R^{-p} \int_0^{2\pi} \log |h(re^{i\phi}\mathfrak{a})| e^{-ip\phi}\, d\phi$$

$$- \frac{1}{p} \sum_{|u|<R} \nu(u; \mathfrak{a}) u^{-p} + \frac{1}{p} \sum_{|u|<R} \nu(r; \mathfrak{a}) (\bar{u}/R^2)^p.$$

Proposition 6.5 implies

$$\log h(z\mathfrak{a}) = \sum_{u \neq 0} \nu(u; \mathfrak{a}) \left[\log\left(1 - \frac{z}{u}\right) + \sum_{p=1}^{q} \frac{1}{p}\left(\frac{z}{u}\right)^p \right]$$

$$= - \sum_{p=q+1}^{\infty} \frac{1}{p} \left(\sum_{u \neq 0} \nu(u, \mathfrak{a}) u^{-p} \right) z^p$$

for $z \in \mathbf{C}(s)$. Therefore

$$H_p(\mathfrak{a}) = -\frac{1}{p}\sum_{u\neq 0} \nu(u;\mathfrak{a})u^{-p}$$

for $p \geqslant q+1$. Hence, if $p \geqslant q+1$, then

$$\frac{1}{\pi}R^{-p}\int_0^{2\pi} \log|h(Re^{i\phi}\mathfrak{a})|\, e^{-ip\phi}\, d\phi$$
$$= -\frac{1}{p}\sum_{|u|\geqslant R} \nu(u;\mathfrak{a})u^{-p} - \frac{1}{p}\sum_{|u|<R} \nu(u;\mathfrak{a})\, (\bar{u}/R)^p$$

which implies

$$\left|\frac{1}{\pi}R^{-p}\int_0^{2\pi} \log|h(Re^{i\phi}\mathfrak{a})|\, e^{-ip\phi}\, d\phi\right|$$
$$\leqslant \frac{1}{p}\sum_{|u|\geqslant R} \nu(u;\mathfrak{a})\,|u|^{-p} + \frac{1}{p}\sum_{|u|\leqslant R} \nu(u;\mathfrak{a})\, (|u|/R^2)^p$$
$$= -\frac{1}{p}n_\nu(R;\mathfrak{a})R^{-p} + \int_R^\infty n_\nu(t;\mathfrak{a})t^{-p-1}\, dt$$
$$+ \frac{1}{p}n_\nu(R;\mathfrak{a})R^{-p} - \int_R^\infty n_\nu(t;\mathfrak{a})\, t^{p-1}\, dt\, R^{-2p}$$
$$\leqslant \int_R^\infty n_\nu(t;\mathfrak{a})t^{-p-1}\, dt$$

for all $p \geqslant q+1$ and all $\mathfrak{a}\in T_{\mathfrak{x}}$. If $\mathfrak{z}\in P_{\mathfrak{x}}(\cdot r \cdot)$ and $\mathfrak{a}\in T_{\mathfrak{x}}$, then

$$|Q_p(\mathfrak{z}/\mathfrak{a})| \leqslant \binom{p+m}{p}\|\mathfrak{z}/\mathfrak{a}\|^p \leqslant \binom{p+m}{p}r^p.$$

Therefore

$$\left|\int_{T_{\mathfrak{x}}} \log|h(R\mathfrak{y})|\, Q_p(\mathfrak{z}/(R\mathfrak{y}))\Omega(\mathfrak{y})\right|$$
$$= \left|\int_{T_{\mathfrak{x}}} \frac{1}{2\pi}R^{-p}\int_0^{2\pi} \log|h(Re^{i\phi}\mathfrak{a})|\, e^{-ip\phi}\, d\phi\, Q_p(\mathfrak{z}/\mathfrak{a})\,\Omega(\mathfrak{a})\right|$$
$$\leqslant \binom{p+m}{p}r^p\int_R^\infty n_{\nu\mathfrak{x}}(t)\,t^{-p-1}\, dt$$
$$\leqslant \binom{p+m}{p}r^p R^{q+1-p}\int_R^\infty n_{\nu\mathfrak{x}}(t)\,t^{-q-2}\, dt$$
$$\leqslant \binom{p+m}{p}\left(\frac{1}{2}\right)^p (2r)^{q+1}\int_R^\infty n_{\nu\mathfrak{x}}(t)\,t^{-q-2}\, dt$$

for $p \geqslant q+1$. Hence, if $\mathfrak{z}\in P_{\mathfrak{x}}(\cdot r\cdot)$, then

$$\left|\int_{T_{\mathfrak{x}}} \log|h(R\mathfrak{y})|\, \Lambda'(\mathfrak{z}/(R\mathfrak{y}),q)\Omega(\mathfrak{y})\right|$$
$$= \left|-\sum_{p=q+1}\int_{T_{\mathfrak{x}}} \log|h(R\mathfrak{y})|\, Q_p(\mathfrak{z}/(R\mathfrak{y}))\Omega(\mathfrak{y})\right|$$
$$\leqslant 2^m(2r)^{q+1}\int_R^\infty n_{\nu\mathfrak{x}}(t)\,t^{-q-2}\, dt \longrightarrow 0$$

for $R\to\infty$ because $q+1$ is an exponent of convergence. q.e.d.

Now, Ronkin's integral representation of the canonical function can be obtained easily. The reader is urged to compare the method of Ronkin with the method which was employed here. (See Ronkin [35, pp. 127–137].) In both instances the reader will notice an affinity to the power series method of Kneser and the Fourier series method of Rubel-Taylor-Kujala for functions of finite λ-type.

THEOREM 10.19 (RONKIN [35]). *Take $v \geqslant 0$ to be a nonnegative divisor on $\mathbf{C}^m$. Define $A = \operatorname{supp} v$. Assume $0 \notin A \neq \varnothing$. Let $q + 1 \in \mathbf{N}$ be an exponent of convergence of v. Let h be the canonical function of v for the constant weight q. Take $\mathfrak{x} \in \mathbf{R}^{+m}$ with $|\mathfrak{x}| = 1$. Take s maximal with $A \cap P_{\mathfrak{x}}(\cdot s \cdot) = \varnothing$. Then the function h is also the canonical function of v for $\mathfrak{x}$ and for the constant weight function q. If $\mathfrak{z} \in P_{\mathfrak{x}}(\cdot s \cdot)$, then $h(\mathfrak{z}) \neq 0$ and*

$$\log h(\mathfrak{z}) = \int_{A\{\mathfrak{x}\}} v(\mathfrak{y}) \Lambda(\mathfrak{z}/\mathfrak{y}, q) \xi(\mathfrak{y}).$$

PROOF. Let f be the canonical function of v for $\mathfrak{x}$ and for the constant weight function q. Take $s < R < \infty$ and $\mathfrak{z} \in P_{\mathfrak{x}}(\cdot s \cdot)$. The Jensen-Poisson formula implies

$$\log h(\mathfrak{z}) = -2 \int_{T_{\mathfrak{x}}} \log |h(R\mathfrak{y})| \Lambda'(\mathfrak{z}/R\mathfrak{y}, q) \Omega(\mathfrak{y})$$
$$+ \int_{A(\mathfrak{x}, R)} v(\mathfrak{y}) \Lambda(\mathfrak{z}/\mathfrak{y}, q) \xi(\mathfrak{y})$$
$$- \int_{A(\mathfrak{x}, R)} v(\mathfrak{y}) \Lambda(\mathfrak{z}\overline{\mathfrak{y}}/(R^2\mathfrak{r}^2), q) \xi(\mathfrak{y}).$$

If $\mathfrak{z} \in P_{\mathfrak{x}}(\cdot s \cdot)$ and $\mathfrak{y} \in A(\mathfrak{x}, R)$, then $\|\mathfrak{z}\overline{\mathfrak{y}}/(R^2\mathfrak{x}^2)\| \leqslant s/R < 1$. Hence

$$|\Lambda(\mathfrak{z}\overline{\mathfrak{y}}/(R^2\mathfrak{x}^2), q)| \leqslant (q + 2)^m (1 - s/R)^{-m} (s/R)^{q+1}$$

and

$$\left| \int_{A(\mathfrak{x}, R)} v(\mathfrak{y}) \Lambda(\mathfrak{z}\overline{\mathfrak{y}}/(R^2\mathfrak{x}^2), q) \xi(\mathfrak{y}) \right| \leqslant (n_{v\mathfrak{x}}(R)/R^{q+1}) (q + 2)^m (1 - s/R)^{-m} s^{q+1} \to 0$$

for $R \to \infty$. Together with Lemma 10.18 this implies

$$\log h(\mathfrak{z}) = \int_{A\{\mathfrak{x}\}} v(\mathfrak{y}) \Lambda(\mathfrak{z}/(\mathfrak{y}, q)) \xi(\mathfrak{y}) = \log f(\mathfrak{z})$$

for all $\mathfrak{z} \in P_{\mathfrak{x}}(\cdot s \cdot)$. Therefore $f = h$ on $\mathbf{C}^m$. q.e.d.

Now, divisors on a polydisc shall be considered. Although the introduction of weights would be quite natural, only the situation without weights shall be studied here.

Take $\mathfrak{x} \in \mathbf{R}^{+m}$ with $|\mathfrak{x}| = 1$. Let v be a nonnegative divisor on $P_{\mathfrak{x}}$. Then v is said to satisfy a Blaschke condition on $P_{\mathfrak{x}}$ if and only if

$$\int_0^1 n_{v\mathfrak{x}}(t) \, dt < +\infty.$$

If $0 \notin A = \operatorname{supp} v$, this is equivalent to $N_{v\mathfrak{x}}(1) < \infty$, respectively to

$$\int_{A\{\mathfrak{x}\}} v(1 - \tau_0) \xi < +\infty.$$

LEMMA 10.20. *Take $\mathfrak{x} \in \mathbf{R}_+^m$ with $|\mathfrak{x}| = 1$. Take $0 < \theta < 1$ and $0 < r < R < 1$. Take $\mathfrak{z} \in P_{\mathfrak{x}}(\cdot \theta r \cdot)$ and $\mathfrak{y} \in \widetilde{T}_{\mathfrak{x}}$ with $r \leqslant |\mathfrak{y}| \leqslant R$. Then*

$$|\Lambda(\mathfrak{z}/\mathfrak{y}) - \Lambda(\mathfrak{z}\overline{\mathfrak{y}}/(R^2 r^2))| \leqslant 2(1 - \theta)^{-m} (1 - |\mathfrak{y}|).$$

PROOF. Here $\mathfrak{z} = (z_1, \cdots, z_m)$ with $|z_\mu| \leqslant \theta r x_\mu$ for $\mu = 1, \cdots, m$. Also

$\mathfrak{y} = (y_1, \cdots, y_m) = u\mathfrak{a}$ with $u \in \mathbf{C}$ and $\mathfrak{a} = (a_1, \cdots, a_m) \in T_{\mathfrak{r}}$. Hence $|\mathfrak{a}| = |\mathfrak{r}| = 1$ and $r \leqslant |\mathfrak{y}| = |u| \leqslant R$. Also $y_\mu = ua_\mu$ and $\mathfrak{a}\bar{\mathfrak{a}} = \mathfrak{r}\mathfrak{r}$. Therefore

$$\left\| \frac{\mathfrak{z}}{\mathfrak{y}} \right\| = \operatorname{Max} \{|z_\mu|/|ua_\mu| \mid \mu = 1, \cdots, m\} \leqslant \theta < 1,$$

$$\mathfrak{z}\bar{\mathfrak{y}}/(R^2\mathfrak{r}^2) = \mathfrak{z}\bar{u}\bar{\mathfrak{a}}/(R^2\mathfrak{a}\bar{\mathfrak{a}}) = (\mathfrak{z}/\mathfrak{y})\,(|u|/R)^2,$$

$$\|\mathfrak{z}\bar{\mathfrak{y}}/(R^2\mathfrak{r}^2)\| = \|\mathfrak{z}/\mathfrak{y}\|\,(|u|/R)^2 \leqslant \theta < 1;$$

Take $p \in \mathbf{N}$, then

$$\frac{R^{2p} - |u|^{2p}}{R^{2p}} \leqslant 1 - |u|^{2p} \leqslant 2p(1 - |u|) = 2p(1 - |\mathfrak{y}|).$$

Also

$$|Q_p(\mathfrak{z}/\mathfrak{y})| \leqslant \binom{m+p}{p} \|\mathfrak{z}/\mathfrak{y}\|^p \leqslant \binom{m+p}{p} \theta^p.$$

Therefore

$$\frac{1}{p}\,|Q_p(\mathfrak{z}/\mathfrak{y}) - Q_p(\mathfrak{z}\bar{\mathfrak{y}}/(R^2\mathfrak{r}^2))| \leqslant \frac{1}{p}\,|Q_p(\mathfrak{z}/\mathfrak{y})|\,|1 - (|u|/R)^{2p}|$$

$$\leqslant 2\binom{p+m}{p}\theta^p(1 - |\mathfrak{y}|).$$

Consequently

$$\left|\Lambda\left(\frac{\mathfrak{z}}{\mathfrak{y}}\right) - \Lambda\left(\frac{\mathfrak{z}\bar{\mathfrak{y}}}{R^2\mathfrak{r}^2}\right)\right| = \left|\sum_{p=1}^{\infty} \frac{1}{p}\left(Q_p\left(\frac{\mathfrak{z}}{\mathfrak{y}}\right) - Q_p\left(\frac{\mathfrak{z}\bar{\mathfrak{y}}}{R^2\mathfrak{r}^2}\right)\right)\right|$$

$$\leqslant 2\sum_{p=1}^{\infty}\binom{p+m}{p}\theta^p\,(1 - |\mathfrak{y}|)$$

$$= 2(1-\theta)^{-m}\,(1 - |\mathfrak{y}|). \quad \text{q.e.d.}$$

LEMMA 10.21. *Take $\mathfrak{r} \in \mathbf{R}^{+m}$ with $|\mathfrak{r}| = 1$. Let $\nu \geqslant 0$ be a nonnegative divisor on $P_{\mathfrak{r}}$. Define $A = \operatorname{supp} \nu$. Assume $0 \notin A \neq \varnothing$. Take $0 < r < R < 1$. Define $B = A(\mathfrak{r}, R) - A(\mathfrak{r}, r)$ and $C = A\{\mathfrak{r}\} - A(\mathfrak{r}, r)$. Assume that ν satisfies a Blaschke condition on $P_{\mathfrak{r}}$. Then, for each $\mathfrak{z} \in P_{\mathfrak{r}}(\cdot r \cdot)$, the integrals*

$$D_r^R(\mathfrak{z}) = \int_B \nu(\mathfrak{y})\,[\Lambda(\mathfrak{z}/\mathfrak{y}) - \Lambda(\mathfrak{z}\bar{\mathfrak{y}}/(R^2\mathfrak{r}^2))]\,\xi(\mathfrak{y}),$$

$$D_r(\mathfrak{z}) = \int_C \nu(\mathfrak{y})\,[\Lambda(\mathfrak{z}/\mathfrak{y}) - \Lambda(\mathfrak{z}\bar{\mathfrak{y}}/\mathfrak{r}^2)]\,\xi(\mathfrak{y})$$

exist and define holomorphic functions D_r^R and D_r on $P_{\mathfrak{r}}(\cdot r \cdot)$. Moreover, $D_r^R \to D_r$ for $R \to 1$ uniformly on each compact subset of $P_{\mathfrak{r}}(\cdot r \cdot)$.

PROOF. Obviously, D_r^R is defined and holomorphic on $P_{\mathfrak{r}}(r)$. Define

$$\chi_r(\mathfrak{y}) = \begin{cases} 1 & \text{if } \mathfrak{y} \in A(\mathfrak{r}, R), \\ 0 & \text{if } \mathfrak{y} \in A\{\mathfrak{r}\} - A(\mathfrak{r}, R). \end{cases}$$

Take $0 < \theta < 1$. Take $\mathfrak{z} \in P_{\mathfrak{r}}(\cdot \theta r \cdot)$. Take $\mathfrak{y} \in C$. Then

$$\chi_r(\mathfrak{y})\,[\Lambda(\mathfrak{z}/\mathfrak{y}) - \Lambda(\mathfrak{z}\bar{\mathfrak{y}}/(R^2\mathfrak{r}^2))] \longrightarrow \Lambda(\mathfrak{z}/\mathfrak{y}) - \Lambda(\mathfrak{z}\bar{\mathfrak{y}}/\mathfrak{r}^2)$$

for $R \to 1$. Lemma 10.20 implies

$$|\chi_r(\mathfrak{y})\ [\Lambda(\mathfrak{z}/\mathfrak{y}) - \Lambda(\mathfrak{z}\bar{\mathfrak{y}}/(R^2\mathfrak{r}^2))]| \leqslant 2(1-\theta)^{-m}(1-|\mathfrak{y}|)$$

where $\int_C \nu(\mathfrak{y})\,(1-|\mathfrak{y}|)\,\xi(\mathfrak{y}) < \infty$ by the Blaschke condition. The Lebesgue bounded convergence theorem implies the existence of D_r on $P_\mathfrak{r}(\cdot\theta r\cdot)$ and the uniform convergence of $D_r^R \to D_r$ for $R \to 1$ on $P_\mathfrak{r}(\cdot\theta r\cdot)$ for each $\theta \in \mathbf{R}(0, 1)$. Especially, D_r is defined and holomorphic on $P_\mathfrak{r}(\cdot r\cdot)$ and $D_r^R \to D_r$ for $R \to 1$ uniformly on each compact subset of $P_\mathfrak{r}(\cdot r\cdot)$. q.e.d.

Now, the Blaschke integral of a divisor on a polydisc satisfying a Blaschke condition can be introduced.

THEOREM 10.22. BLASCHKE INTEGRAL THEOREM. *Take $\mathfrak{r} \in \mathbf{R}^{+m}$ with $|\mathfrak{r}| = 1$. Let $\nu \geqslant 0$ be a nonnegative divisor on $P_\mathfrak{r}$. Define $A = \operatorname{supp}\nu$. Assume $0 \notin A \neq \varnothing$. Let s be maximal with $A \cap P(\cdot s\cdot) = \varnothing$. Assume that ν satisfies a Blaschke condition on $P_\mathfrak{r}$. Then there exists a holomorphic function B_ν on $P_\mathfrak{r}$ with $B_\nu(0) = 1$ and $\mu_{B_\nu} = \nu$ such that $B_\nu(\mathfrak{z}) \neq 0$ for $\mathfrak{z} \in P_\mathfrak{r}(\cdot s\cdot)$ and such that*

$$\log B_\nu(\mathfrak{z}) = \int_{A\{\mathfrak{r}\}} \nu(\mathfrak{y})\ [\Lambda(\mathfrak{z}/\mathfrak{y}) - \Lambda(\mathfrak{z}\bar{\mathfrak{y}}/(\mathfrak{r}\mathfrak{r}))]\,\xi(\mathfrak{y}) \tag{10.5}$$

for all $\mathfrak{z} \in P_\mathfrak{r}(\cdot s\cdot)$. The function B_ν is uniquely defined by ν and is called the Blaschke function of ν on $P_\mathfrak{r}$.

PROOF. Take any $r \in \mathbf{R}[s, 1)$. Define D_r as in Lemma 10.21. Then $D_r(0) = 0$. Take $\mathfrak{z} \in P_\mathfrak{r}(\cdot r\cdot)$ and $\mathfrak{y} \in A(\mathfrak{r}, r)$. Then $\mathfrak{y} = u\mathfrak{a}$ with $u \in \mathbf{C}(r)$ and $\mathfrak{a} \in T_\mathfrak{r}$. Then

$$\|\mathfrak{z}\bar{\mathfrak{y}}/(\mathfrak{r}\mathfrak{r})\| = \|\mathfrak{z}\bar{u}\,\bar{\mathfrak{a}}/(\mathfrak{a}\bar{\mathfrak{a}})\| = \|\mathfrak{z}\bar{u}/\mathfrak{a}\| \leqslant r^2 < 1.$$

Hence a holomorphic function E_r with $E_r(0) = 0$ is defined on $P_\mathfrak{r}(\cdot r\cdot)$ by

$$E_r(\mathfrak{z}) = \int_{A(\mathfrak{r},r)} \nu(\mathfrak{y})\,\Lambda(\mathfrak{z}\bar{\mathfrak{y}}/(\mathfrak{r}\mathfrak{r}))\,\xi(\mathfrak{y})$$

for all $\mathfrak{z} \in P_\mathfrak{r}(\cdot r\cdot)$. According to Theorem 10.14 a holomorphic function F_r on $P_\mathfrak{r}(\cdot r\cdot)$ with $F_r(0) = 1$ and $\mu_{F_r} = \nu\,|P_\mathfrak{r}(\cdot r\cdot)$ exists such that $F_r(\mathfrak{z}) \neq 0$ for all $\mathfrak{z} \in P_\mathfrak{r}(\cdot s\cdot)$ and such that

$$\log F_r(\mathfrak{z}) = \int_{A(\mathfrak{r},r)} \nu(\mathfrak{y})\,\Lambda(\mathfrak{z}/\mathfrak{y})\,\xi(\mathfrak{y})$$

for all $\mathfrak{z} \in P_\mathfrak{r}(\cdot s\cdot)$. The function $B_\nu^r = F_r \exp(D_r - E_r)$ is holomorphic on $P_\mathfrak{r}(\cdot r\cdot)$ with $B_\nu^r(0) = 1$. If $\mathfrak{z} \in P_\mathfrak{r}(\cdot s\cdot)$, then

$$\log B_\nu^r(\mathfrak{z}) = \int_{A\{\mathfrak{r}\}} \nu(\mathfrak{y})\ [\Lambda(\mathfrak{z}/\mathfrak{y}) - \Lambda(\mathfrak{z}\bar{\mathfrak{y}}/(\mathfrak{r}\mathfrak{r}))]\,\xi(\mathfrak{y}).$$

Therefore B_ν^r does not depend on $r \in \mathbf{R}[s, 1)$. One and only one holomorphic function B_ν on $P_\mathfrak{r}$ exists such that $B_\nu\,|P_\mathfrak{r}(\cdot r\cdot) = B_\nu^r$ for each $r \in \mathbf{R}[s, 1)$. Obviously $B_r(0) = 1$ and $\mu_{B_\nu} = \nu$ and (10.5) holds on $P_\mathfrak{r}(\cdot s\cdot)$. q.e.d.

Unfortunately, there are no means at present to estimate B_ν. Suppose that ν is a nonnegative divisor on $P_\mathfrak{r}$ with $\mathfrak{r} \in \mathbf{R}^{+m}$ and $|\mathfrak{r}| = 1$. Assume that there exists an open neighborhood of $T_\mathfrak{r}$ in $\mathbf{C}^m$ such that $U \cap \operatorname{supp}\nu = \varnothing$. Then ν satisfies a Blaschke condition and the Blaschke function B_ν exists. According to a theorem of Rudin [37], the divisor ν is the divisor of a bounded holomorphic function. Is the Blaschke function B_ν bounded under these circumstances?

Now, bounded holomorphic functions on the polydisc $P_\mathfrak{r}$ shall be considered. The following lemma is a trivial consequence of the Jensen formula.

LEMMA 10.23. *Take $\mathfrak{r} \in \mathbf{R}^{+m}$ with $|\mathfrak{r}| = 1$. Let f be a bounded holomorphic function on $P_\mathfrak{r}$ with $f(0) \neq 0$. Then $\nu = \mu_f$ satisfies a Blaschke condition on $P_\mathfrak{r}$. Moreover, if $\mathfrak{a} \in T_\mathfrak{r}$ then*

$$N_\nu(1;\mathfrak{a}) \leqslant \log M_f(1) - \log|f(0)|,$$

$$N_{\nu\mathfrak{r}}(1) \leqslant \log M_f(1) - \log|f(0)|.$$

Let f be a bounded, holomorphic function on P, then the limit

$$f^*(\mathfrak{a}) = \lim_{0<r\to 1} f(r\mathfrak{a})$$

exists for almost all $\mathfrak{a} \in T_\mathfrak{r}$. (For instance, see Rudin [38, Theorem 3.3.3].)

LEMMA 10.24. *Take $\mathfrak{r} \in \mathbf{R}^{+m}$ with $|\mathfrak{r}| = 1$. Let f be a bounded, holomorphic function on $P_\mathfrak{r}$ with $f(0) \neq 0$. Define $\nu = \mu_f$. Then $(\log|f^*|)\Omega$ is integrable over $T_\mathfrak{r}$ with*

$$N_{\nu\mathfrak{r}}(1) + \log|f(0)| \leqslant \int_{T_\mathfrak{r}} \log|f^*|\Omega. \tag{10.6}$$

PROOF. Define $M = M_f(1) > 0$. Take $0 < r < 1$. Then $\log(M/|f(r\mathfrak{y})|) \geqslant 0$ for all $\mathfrak{y} \in T_\mathfrak{r}$ and

$$\log\frac{M}{|f(0)|} - N_{\nu\mathfrak{r}}(r) = \int_{T_\mathfrak{r}} \log\frac{M}{|f(r\mathfrak{y})|}\,\Omega(\mathfrak{y}).$$

The integral decreases with r. Fatou's lemma implies

$$0 \leqslant \int_{T_\mathfrak{r}} \log\frac{M}{|f^*|}\Omega(\mathfrak{y}) \leqslant \log\frac{M}{|f(0)|} - N_{\nu\mathfrak{r}}(1)$$

which implies (10.6) immediately. q. e. d.

Again take $\mathfrak{r} \in \mathbf{R}^{+m}$ with $|\mathfrak{r}| = 1$. A bounded, holomorphic function f on $P_\mathfrak{r}$ is said to be good if and only if

$$\int_{T_\mathfrak{r}} |\log|f(r\mathfrak{y})/f(\mathfrak{y})||\,\Omega(\mathfrak{y}) \longrightarrow 0$$

for $r \to 1$. This definition deviates from Rudin's definition in [26]. Rudin's good functions are the good, inner functions introduced below.

THEOREM 10.25. FACTORIZATION. *Take $\mathfrak{r} \in \mathbf{R}^{+m}$ with $|\mathfrak{r}| = 1$. Let f be a good, holomorphic function on $P_\mathfrak{r}$ with $f(0) > 0$. Define $\nu = \mu_f$. Then ν satisfies a Blaschke condition on $P_\mathfrak{r}$. Let B_ν be the Blaschke function for ν. Then*

$$N_{\nu\mathfrak{r}}(1) + \log f(0) = \int_{T_\mathfrak{r}} \log|f^*|\Omega \tag{10.7}$$

for each $\mathfrak{z} \in P_\mathfrak{r}$, the integral

$$F(\mathfrak{z}) = \int_{T_\mathfrak{r}} \log|f^*(\mathfrak{y})|\,[1 - 2\Lambda'(\mathfrak{z}/\mathfrak{y})]\,\Omega(\mathfrak{y})$$

exists and defines a holomorphic function on $P_{\mathfrak{x}}$. *Moreover*

$$f = B_\nu \exp (F - N_{\nu \mathfrak{x}}(1)).$$

(*If* $\nu \equiv 0$, *then* $B_\nu \equiv 1$.)

PROOF. Because f is good,

$$N_{\nu \mathfrak{x}}(r) + \log f(0) = \int_{T_{\mathfrak{x}}} \log |f(r\mathfrak{y})| \, \Omega(\mathfrak{y}) \longrightarrow \int_{T_{\mathfrak{x}}} \log |f^*| \, \Omega$$

for $r \to 1$ which implies (10.7). Define $A = \operatorname{supp} \nu$. If $\nu \equiv 0$ define $B_\nu \equiv 1$. If $\nu \not\equiv 0$, then $0 \in A \neq \emptyset$. Take s maximal such that $A \cap P \, (\cdot s \cdot) = \emptyset$. Cauchy's integral formula implies

$$1 = \int_{T_{\mathfrak{x}}} [1 - \Lambda'(\mathfrak{z}/\mathfrak{y})] \, \Omega(\mathfrak{y});$$

hence

$$\int_{T_{\mathfrak{x}}} \Lambda'(\mathfrak{z}/\mathfrak{y}) \Omega(\mathfrak{y}) = 0$$

for all $\mathfrak{z} \in P_{\mathfrak{x}}$. Take $R \in \mathbf{R}(s, 1)$. The Jensen-Poisson formula of Theorem 10.11 for $\mathfrak{z} \in P \, (\cdot s \cdot)$ implies

$$\log f(\mathfrak{z}) = \log f(0)$$
$$= 2 \int_{T_{\mathfrak{x}}} \log |f(R\mathfrak{y})| \Lambda'(\mathfrak{z}/\mathfrak{y}) \, \Omega(\mathfrak{y})$$
$$+ \int_{A(\mathfrak{x}, R)} \nu(\mathfrak{y}) \, [\Lambda(\mathfrak{z}/\mathfrak{y}) - \Lambda(\mathfrak{z}\bar{\mathfrak{y}} /(R^2 \mathfrak{x}^2))] \xi(\mathfrak{y}) \, .$$

Because f is good,

$$\int_{T_{\mathfrak{x}}} \log |f(R\mathfrak{y})| \Lambda'(\mathfrak{z}/\mathfrak{y}) \Omega(\mathfrak{y}) \longrightarrow \int_{T_{\mathfrak{x}}} \log |f^*(\mathfrak{y})| \Lambda'(\mathfrak{z}/\mathfrak{y}) \Omega(\mathfrak{y})$$

for $R \to 1$. Especially, F exists and is holomorphic on $P_{\mathfrak{x}}$. (The limit is uniform on compact subsets of $P_{\mathfrak{x}}$.) Lemma 10.21 with $r = s$ implies

$$\log f = \log f(0) - \int_{T_{\mathfrak{x}}} \log |f^*| \, \Omega + F + \log B_\nu$$
$$= - N_{\nu \mathfrak{x}}(1) + F + \log B_\nu$$

on $P_{\mathfrak{x}}(\cdot s \cdot)$. Hence $f = B_\nu \exp (F - N_{\nu \mathfrak{x}}(1))$ on $P_{\mathfrak{x}}(\cdot s \cdot)$ and consequently on $P_{\mathfrak{x}}$. q. e. d.

Take $\mathfrak{x} \in \mathbf{R}^{+m}$ with $|\mathfrak{x}| = 1$. A bounded, holomorphic function f on $P_{\mathfrak{x}}$ is said to be an inner function on $P_{\mathfrak{x}}$ if and only if $f^*(\mathfrak{z}) = 1$ for almost all $\mathfrak{z} \in T_{\mathfrak{x}}$. If f is an inner function, then $|f(\mathfrak{z})| \leq 1$ for all $\mathfrak{z} \in P_{\mathfrak{x}}$. Therefore an inner function f is good if and only if

$$\int_{T_{\mathfrak{x}}} \log |f(r\mathfrak{y})| \, \Omega(\mathfrak{y}) \longrightarrow 0$$

for $r \to 1$; equivalently, an inner function f with $f(0) \neq 0$ is good if and only if

$$\log |f(0)| = - N_{\nu \mathfrak{x}}(1).$$

Now, Theorem 10.25 implies the following theorem easily.

THEOREM 10.26. *Take* $\mathfrak{x} \in \mathbf{R}^{+m}$ *with* $|\mathfrak{x}| = 1$. *Let* f *be a good inner function on* $P_{\mathfrak{r}}$ *with* $f(0) > 0$. *Define* $\nu = \mu_f$ *and* $A = \operatorname{supp} \nu$. *Assume* $A \neq \emptyset$. *Take* s *maximal with* $A \cap P_{\mathfrak{r}}(\cdot s \cdot) = \emptyset$. *Then* ν *satisfies a Blaschke condition. Let* B_ν *be the Blaschke function of* ν *on* $P_{\mathfrak{r}}$. *Then*

$$f = B_\nu \exp(-N_{\nu\mathfrak{r}}(1)).$$

If $\mathfrak{z} \in P(\cdot s \cdot)$, *then*

$$\log f(\mathfrak{z}) = -N_{\nu\mathfrak{r}}(1) + \int_{A\{\mathfrak{r}\}} \nu(\mathfrak{y})\, [\Lambda(\mathfrak{z}/\mathfrak{y}) - \Lambda(\mathfrak{z}\bar{\mathfrak{y}}/(\mathfrak{r}\mathfrak{r}))]\, \xi(\mathfrak{y}) .$$

Thus f is uniquely determined by its divisor. The representation part of Problem 5.4.4 in Rudin's book [38] is solved. The construction part is considerably more difficult and can be reformulated as the question: When is the Blaschke function bounded?

The results of this section are rooted in Ronkin's paper [35, §§1, 2], but go considerably beyond Ronkin's results in that paper. Basically there is an affinity to ideas expounded by Stefan Bergman since about 40 years. He stresses the importance of the distinguished boundary. A direct comparison is difficult, since Bergman has special configurations and it is cumbersome to compare Bergman's terms with those of this section. Bergman considers the case $m = 2$ only. Some of the many papers of Bergman are listed in the references at the end of this report.

11. Entire Functions for Analytic Sets

Let A be an analytic subset of $\mathbf{C}^m$. A well-known result asserts that A is the common zero set of $(m+1)$ entire functions. Can these functions be chosen such that their growth can be estimated in terms of n_A? If A is pure $(m-1)$-dimensional, and has finite order, the question is answered by the construction of the canonical function. If A is pure zero-dimensional, the problem was solved by Pan [33]. In the same paper, Pan solves the problem also in the case where A is pure p-dimensional with $0 < p < m-1$ and where A behaves sufficiently regular at infinity. Skoda [44] solved the whole problem by an ingenious method. Here a short outline of the highlights of Skoda's proof shall be given. Skoda considers nonnegative, closed currents. However this report shall be restricted to analytic sets. Also Skoda derives several types of estimates. This report shall be restricted to one type of estimate. At first, Skoda constructs a pluri-subharmonic function on $\mathbf{C}^m$ which is of class C^∞ on $\mathbf{C}^m - A$ and which becomes $-\infty$ on A of sufficient strength and which can be estimated above in terms of n_A. Then Hörmander's theory is used to construct the entire functions. Assume $m > 1$ for this section.

Consider the construction of the pluri-subharmonic function first. Some concepts and notations shall have to be introduced.

Take $\mathfrak{a} = (a_1, \cdots, a_m) \in \mathbf{C}^m$. The differential operators

$$\nabla_{\mathfrak{a}} = \sum_{q=1}^{m} a_q \partial/\partial z_q, \qquad \overline{\nabla}_{\mathfrak{a}} = \sum_{q=1}^{m} \bar{a}_q \partial/\partial \bar{z}_q$$

are defined. If $V : G \to \mathbf{R}$ is a function of class C^2 on the open subset $G \neq \emptyset$ of $\mathbf{C}^m$, then the hermitian form

$$\nabla_{\mathfrak{a}} \overline{\nabla}_{\mathfrak{a}} V = \sum_{\mu,\nu=1}^{m} a_\mu \bar{a}_\nu \partial^2 V/\partial z_\mu \partial \bar{z}_\nu$$

is called the hessian of V. The function V is pluri-subharmonic, if and only if $\nabla_{\mathfrak{a}} \overline{\nabla}_{\mathfrak{a}} V \geqslant 0$ for each $\mathfrak{a} \in \mathbf{C}^m$.

Define $\mathfrak{T}(p, m)$ as the set of all increasing, injective maps $\nu : \mathbf{N}[1, p] \to \mathbf{N}[1, m]$. If $\nu \in \mathfrak{T}(p, m)$, define

$$dz_\nu = dz_{\nu(1)} \wedge \cdots \wedge dz_{\nu(p)}.$$

A differential form χ of bidegree (p, q) on $\mathbf{C}^m$ can be written as

$$\chi = \sum_{\mu \in \mathfrak{T}(p,m)} \sum_{\nu \in \mathfrak{T}(q,m)} \chi_{\mu\nu} dz_\mu \wedge d\bar{z}_\nu$$

where $\chi_{\mu\nu}$ are functions on $\mathbf{C}^m$. Define

$$|\chi(\mathfrak{z})| = \Big(\sum_{\mu \in \mathfrak{T}(p,m)} \sum_{\nu \in \mathfrak{T}(q,m)} |\chi_{\mu\nu}(\mathfrak{z})|^2 \Big)^{1/2}.$$

At first consider the case where the given analytic set A has pure dimension p with $0 < p < m$. One and only one effective analytic chain ν_a exists such that $\nu_A(\mathfrak{z}) = 1$ for all $\mathfrak{z} \in \mathfrak{R}(A)$. Define

$$n_A = n_{\nu_A}, \qquad N_A = N_{\nu_A}.$$

Now, the first step can be taken. Let $\eta \geqslant 0$ be a nonnegative function of class C^∞ with compact support in $\mathbf{C}^m$. A function P of class C^∞ is defined on $\mathbf{C}^m - A \cap \operatorname{supp} \eta$ by

$$P(\mathfrak{z}) = -\int_A \eta |\mathfrak{z} - \mathfrak{x}|^{-2p} v^p(\mathfrak{x})$$

If $\mathfrak{z}_0 \in A$ and $\eta(\mathfrak{z}_0) > 0$, then $P(\mathfrak{z}) \to -\infty$ for $\mathfrak{z} \to \mathfrak{z}_0$. Skoda shows the existence of a constant $c(p, m) > 0$ depending on the dimensions p and m only such that

$$(\nabla_{\mathfrak{a}} \overline{\nabla}_{\mathfrak{a}} P)(\mathfrak{z}) \geqslant -c(p, m)\, |\mathfrak{a}|^2 \int_A (|\bar{\partial}\eta| + |\mathfrak{z} - \mathfrak{x}|\, |\partial\bar{\partial}\eta|) \frac{v^p(\mathfrak{x})}{|\mathfrak{z} - \mathfrak{x}|^{2p+1}}$$

for all $\mathfrak{z} \in \mathbf{C}^m - A \cap \operatorname{supp} \eta$ and all $\mathfrak{a} \in \mathbf{C}^m$. Although P is not pluri-subharmonic itself, the deviation is estimated.

The second step globalizes the procedures. A function $\eta : \mathbf{C}^m \times \mathbf{C}^m \to \mathbf{R}[0, 1]$ of class C^∞ is constructed such that $\eta | G = 1$ for some open neighborhood G of the diagonal D of $\mathbf{C}^m \times \mathbf{C}^m$ and such that $(K \times \mathbf{C}^m) \cap \operatorname{supp} \eta$ is compact for each compact subset of $\mathbf{C}^m$. A function $P \leqslant 0$ of class C^∞ is defined on $\mathbf{C}^m - \{0\}$ by

$$P(\mathfrak{z}) = -\int_A \eta(\mathfrak{z}, \mathfrak{x}) |\mathfrak{z} - \mathfrak{x}|^{-2p} v^p(\mathfrak{x})$$

for all $\mathfrak{z} \in \mathbf{C}^m - A$.

Skoda does not take just any such function η but makes two particular choices which lead to different solutions with different estimates. Here, only one of these choices shall be considered.

Take $\epsilon > 0$. Take a decreasing function $\chi : \mathbf{R} \to \mathbf{R}[0, 1]$ of class C^∞ such that $\chi(x) = 1$ if $x \leqslant 1$ and $\chi(x) = 0$ if $x \geqslant 1 + \epsilon$. For each $q \in \mathbf{N}$, define $\chi_q : \mathbf{C}^m \to \mathbf{R}[0, 1]$ as a function of class C^∞ by

$$\chi_q(\mathfrak{z}) = \chi(|\mathfrak{z}|/q)$$

for all $\mathbf{C}^m$. Then

$$\chi_q(\mathfrak{z}) = 1 \quad \text{if } |\mathfrak{z}| \leqslant q,$$
$$\chi_q(\mathfrak{z}) = 0 \quad \text{if } |\mathfrak{z}| \geqslant q(1 + \epsilon),$$
$$\chi_{q+1} \geqslant \chi_q,$$

for all $q \in \mathbf{N}$. Define $\rho_1 = \chi_1$ and $\rho_q = \chi_q - \chi_{q-1}$ for $q \geqslant 1$. Then ρ_q is a function of class C^∞ with $0 \leqslant \rho_q \leqslant 1$ such that $\Sigma_{q=1}^\infty \rho_q = 1$. Also $\rho_q(\mathfrak{z}) \neq 0$ implies $q - 1 \leqslant |\mathfrak{z}| \leqslant q(1 + \epsilon)$. A function η_q of class C^∞ with $0 \leqslant \eta_q \leqslant 1$ is defined by

$$\eta_q(\mathfrak{x}) = \chi(|\mathfrak{x}|/((1 + 2\epsilon)q)).$$

Then

$$\eta_q(\mathfrak{x}) = 1 \quad \text{if } |\mathfrak{x}| \leqslant (1 + 2\epsilon)q,$$
$$\eta_q(\mathfrak{x}) = 0 \quad \text{if } |\mathfrak{x}| \geqslant (1 + 5\epsilon)q.$$

A function $\eta : \mathbf{C}^m \times \mathbf{C}^m \to \mathbf{R}[0, 1]$ of class C^∞ is defined by

$$\eta(\mathfrak{z}, \mathfrak{x}) = \sum_{q=1}^{\infty} \eta_q(\mathfrak{x}) \rho_q(\mathfrak{z}).$$

Then

$$\eta(\mathfrak{z}, \mathfrak{x}) = 1 \quad \text{if } |\mathfrak{z} - \mathfrak{x}| \leqslant 1,$$
$$\eta(\mathfrak{z}, \mathfrak{x}) = 0 \quad \text{if } |\mathfrak{z} - \mathfrak{x}| \geqslant (1 + |\mathfrak{z}|)(2 + 8\epsilon),$$

with this choice of η, the following estimate is obtained:

(11.1) $$(\nabla_{\mathfrak{a}} \overline{\nabla}_{\mathfrak{a}} P)(\mathfrak{z}) \geqslant -c \frac{|\mathfrak{a}|^2}{1 + |\mathfrak{z}|^2} n_A((1 + 5\epsilon)(1 + |\mathfrak{z}|))$$

for all $\mathfrak{a} \in \mathbf{C}^m$ and all $\mathfrak{z} \in \mathbf{C}^m - A$. Here, the constant $c > 0$ depends on m, p, ϵ and χ only.

Now, the following abbreviation is convenient.

(11.2) $$\mu_A(r, \epsilon) = (1 + (\log(1 + r))^2) n_A((1 + \epsilon)(1 + r))$$

for all $r > 0$ and all $\epsilon > 0$.

As a third step, a function Y of class C^∞ on $\mathbf{C}^m$ is constructed such that

(11.3) $$(\nabla_{\mathfrak{a}} \overline{\nabla}_{\mathfrak{a}} Y)(\mathfrak{z}) \geqslant c \frac{|\mathfrak{a}|^2}{(1 + r^2)} n_A((1 + 5\epsilon)(1 + r)),$$
$$Y(\mathfrak{z}) \leqslant c_1 \mu_A(r, 11\epsilon),$$

for all $\mathfrak{z} \in \mathbf{C}^m$ where $|\mathfrak{z}| = r$. Here c is the same constant as in (11.1). The constant $c_1 > 0$ depends on m, p, ϵ and χ only. The construction of Y is respectively simple since only functions of $|\mathfrak{z}|$ have to be considered.

The addition of (11.1) and (11.3) implies, that the function

(11.4) $$V = (m/p)(Y + P)$$

is pluri-subharmonic and of class C^∞ on $\mathbf{C}^m - A$. Because $P \leqslant 0$, the inequality

$$V(\mathfrak{z}) \leqslant c_1 \mu_A(r, 11\epsilon)$$

holds with $|\mathfrak{z}| = r$. Since V is bounded above on each ball, a theorem of Grauert and Remmert [15] asserts that V is pluri-subharmonic on $\mathbf{C}^m$.

Now, the fourth step can be taken. Let U be any open subset of $\mathbf{C}^m$ such that $U \cap A \neq \varnothing$. Then e^{-V} is not integrable over U.

As the fifth step, the case $\dim A = p = 0$ is considered. A function P of class C^∞ is defined on $\mathbf{C}^m - A$ by

$$P(\mathfrak{z}) = \sum_{\mathfrak{x} \in A} \eta(\mathfrak{z}, \mathfrak{x}) \log \frac{|\mathfrak{z} - \mathfrak{x}|}{1 + |\mathfrak{x}|}$$

for all $\mathfrak{z} \in \mathbf{C}^m - A$. Then there exists a constant $c > 0$ depending on m, ϵ and χ only. Then a pluri-subharmonic function V on $\mathbf{C}^m$ exists such that V is of class C^∞ on

$\mathbf{C}^m - A$. Moreover $mY = V - mP$ is of class C^∞ on $\mathbf{C}^m$. Also

$$V(\mathfrak{z}) \leqslant c_1 \mu_A(r, 11\epsilon)$$

for all $\mathfrak{z} \in \mathbf{C}^m$ where $|\mathfrak{z}| = r$. Here the constant $c_1 > 0$ depends on m, ϵ and χ only. So, for $p = 0$, the only difference to the case $p > 0$ is that P may not be negative. Also e^{-V} is not integrable over any open subset U of $\mathbf{C}^m$ with $U \cap A \neq \varnothing$.

Now, let A be any analytic subset of $\mathbf{C}^m$ with $A \neq \mathbf{C}^m$. Then $A = A_0 \cup A_1 \cup \cdots \cup A_{m-1}$ where A_p is either empty or a pure p-dimensional analytic subset of $\mathbf{C}^m$ such that $A_p \cap A_q$ is nowhere dense in A_p and A_q if $p \neq q$ and if $A_p \neq \varnothing \neq A_q$. Then $A_0, \cdots, A_{m-1}$ are well defined by A and are called the pure dimensional components of A. Define

$$n_A = \sum_{p=0}^{m-1} n_{A_p}.$$

Again, define μ_A by (11.2) for each A_p, take the pluri-subharmonic function V_p as constructed before but take the function as constructed for $\epsilon/11$. Define $V = V_0 + \cdots + V_p$. Then V has the following properties:

(P1) The function V is pluri-subharmonic on $\mathbf{C}^m$.

(P2) The function V is of class C^∞ on $\mathbf{C}^m - A$.

(P3) If $\mathfrak{z} \in \mathbf{C}^m$ and if $|\mathfrak{z}| = r$, then $V(\mathfrak{z}) \leqslant c\mu_A(r, \epsilon)$ where the constant $c > 0$ depends on m, ϵ and χ only.

(P4) If U is open in $\mathbf{C}^m$, if $U \cap A \neq \varnothing$, then e^{-V} is not integrable over U.

In the sixth step, Hörmander's theory is applied to obtain the first entire function. If $h : \mathbf{C}^m \to \mathbf{R}$ is a measurable function on $\mathbf{C}^m$ and if χ is a measurable differential form on $\mathbf{C}^m$ define

$$\|\chi\|_h = \left(\int_{\mathbf{C}^m} |\chi|^2 e^{-h} v^m \right)^{1/2}.$$

Let $\mathfrak{L}_h$ be the vector space of all measurable forms χ on $\mathbf{C}^m$ with $\|\chi\|_h < +\infty$. If $\chi \in \mathfrak{L}_h$ and $\psi \in \mathfrak{L}_h$ then

$$\|\chi + \psi\|_h \leqslant \|\chi\|_h + \|\psi\|_h, \qquad \|\chi \wedge \psi\|_h \leqslant \|\chi\|_h \|\psi\|_h.$$

If $h_1 \leqslant h_2$, then $\|\chi\|_{h_2} \leqslant \|\chi\|_{h_1}$. If $h : \mathbf{R}_+ \to \mathbf{R}$ is a function such that $h \circ \tau_0$ is measurable define $\|\chi\|_h = \|\chi\|_{h \circ \tau_0}$. Then the following theorem holds.

THEOREM 11.1 (SKODA [44]). *Let A be an analytic subset of $\mathbf{C}^m$ with $A \neq \mathbf{C}^m$. Let B be an analytic subset of $\mathbf{C}^m$ with* $\dim B = 0$. *Assume that $A \cap B = \varnothing$. Let $\mathfrak{c} \in B$ be a distinguished point in B. Take a number ϵ with $0 < \epsilon < 1$. Then there exists a holomorphic function F on $\mathbf{C}^m$ such that $F | A = 0$ and $F | B > 0$ and $F(\mathfrak{c}) = 1$ and such that there exists a constant c, depending on m and ϵ only such that the integral*

$$\int_{\mathbf{C}^m} |F|^2 \exp(-c\mu_A(\tau_0, \epsilon) - c\mu_B(\tau_0, \epsilon)) \frac{v^m}{(1 + \tau)^2} < \infty \tag{11.5}$$

exists.

PROOF. Let V_A and V_B be pluri-subharmonic functions on $\mathbf{C}^m$ satisfying conditions (P1)–(P4) for A, respectively B. Define

$$V = V_A + V_B, \qquad \mu = \mu_A(\tau_0, \epsilon) + \mu_B(\tau_0, \epsilon).$$

A constant $c > 0$ exists such that $V \leqslant c\mu$.

For each $\mathfrak{b} \in B$, take $r_{\mathfrak{b}} > 0$ such that $|\mathfrak{z} - \mathfrak{b}| \leqslant r_{\mathfrak{b}}$ implies $\mathfrak{z} \notin A$. Take a function $w_{\mathfrak{b}}$ of class C^∞ on $\mathbf{C}^m$ with $0 \leqslant w_{\mathfrak{b}} \leqslant 1$ such that $w_{\mathfrak{b}}(\mathfrak{x}) = 1$ if $|\mathfrak{x} - \mathfrak{b}| \leqslant \frac{1}{2} r_{\mathfrak{b}}$ and $w_{\mathfrak{b}}(\mathfrak{x}) = 0$ if $|\mathfrak{x} - \mathfrak{b}| \geqslant r_{\mathfrak{b}}$. Then

$$\|w_{\mathfrak{b}}\|_{c\mu} < +\infty, \qquad \|\bar\partial w_{\mathfrak{b}}\|_V < +\infty.$$

Take a sequence $\{\epsilon_{\mathfrak{b}}\}_{\mathfrak{b} \in B}$ of positive numbers $\epsilon_{\mathfrak{b}} > 0$ with $\epsilon_{\mathfrak{c}} = 1$ such that

$$\sum_{\mathfrak{b} \in B} \epsilon_{\mathfrak{b}} \|w_{\mathfrak{b}}\|_{c\mu} < +\infty, \qquad \sum_{\mathfrak{b} \in B} \epsilon_{\mathfrak{b}} \|\bar\partial w_{\mathfrak{b}}\|_V < +\infty.$$

The series $w = \Sigma_{\mathfrak{b} \in B} \epsilon_{\mathfrak{b}} w_{\mathfrak{b}}$ is locally finite and defines a function of class C^∞ on $\mathbf{C}^m$. Obviously

$$\|w\|_{c\mu} \leqslant \sum_{\mathfrak{b} \in B} \epsilon_{\mathfrak{b}} \|w_{\mathfrak{b}}\|_{c\mu} < \infty,$$

$$\|w\|_V \leqslant \sum_{\mathfrak{b} \in B} \epsilon_{\mathfrak{b}} \|\bar\partial w_{\mathfrak{b}}\|_V < \infty.$$

Hörmander [18, Theorem 4.4.2] implies the existence of a function u of class C^∞ on $\mathbf{C}^m$ such that $\bar\partial u = -\bar\partial w$ and

$$\int_{\mathbf{C}^m} |u|^2 e^{-V} \frac{v^m}{(1+\tau)^2} \leqslant \tfrac{1}{2} \|\bar\partial w\|_V^2 < +\infty.$$

Let U be any open subset of $\mathbf{C}^m$ such that $U \cap (A \cup B) \neq \varnothing$. Then e^{-V} is not integrable over U. Therefore

$$u \,|\, (A \cup B) = 0.$$

The function $F = u + w$ is of class C^∞ on $\mathbf{C}^m$ with $\bar\partial F = 0$. Hence F is holomorphic. Obviously

$$F|A = u|A + w|A = 0,$$

$$F(\mathfrak{b}) = u(\mathfrak{b}) + w(\mathfrak{b}) = \epsilon_{\mathfrak{b}} \quad \text{if } \mathfrak{b} \in B,$$

$$F(\mathfrak{c}) = \epsilon_{\mathfrak{c}} = 1.$$

Define $\gamma = c\mu + 2\log(1+\tau)$ and $\sigma = V + 2\log(1+\tau)$. Then

$$\|F\|_\gamma \leqslant \|u\|_\gamma + \|w\|_\gamma \leqslant \|u\|_\sigma + \|w\|_{c\mu}$$
$$\leqslant \|\bar\partial w\|_V + \|w\|_{c\mu} < +\infty.$$

Hence

$$\int_{\mathbf{C}^m} |F| e^{-c\mu} (1+\tau)^{-2} v^m < +\infty, \qquad \text{q.e.d.}$$

Let $h \geqslant 0$ be a nonnegative, continuous increasing function on $\mathbf{R}_+$. Let F be a holomorphic function with $\|F\|_h < +\infty$. Let ϵ be a number with $0 < \epsilon < 1$. Take $r > 0$, then

$$\log M_F(r) \leqslant m \log(1/\epsilon) + \log \|F\|_h + h(r + \epsilon).$$

Therefore the function F of Theorem 11.1 satisfies the following inequality:

(11.6) $$\log M_F(r) \leqslant c\mu_A(r+\epsilon, \epsilon) + c\mu_B(r+\epsilon, \epsilon) + 2\log(1+r^2) + c_0$$

where $c > 0$ is a constant depending on m and ϵ only and where $c_0 > 0$ is a constant depending on m, ϵ, A and B.

Now, suppose that an analytic subset A of $\mathbf{C}^m$ with $A \neq \mathbf{C}^m$ is given. Assume that n_A is unbounded. Then take any analytic subset B_1 of $\mathbf{C}^m$ with $\dim B_1 \neq 0$ and $B_1 \cap A = \varnothing$. Since n_A is unbounded, B_1 can be taken such that $n_{B_1} \leqslant n_A$. Therefore $\mu_{B_1} \leqslant \mu_A$. Hence an entire function F_1 on $\mathbf{C}^m$ exists such that $F_1 | A = 0$ and $F | B_1 > 0$ and

$$\log M_{F_1}(r) \leqslant c_1\mu_A(r+\epsilon, \epsilon) + 2\log(1+r^2) + c_1'$$

for all $r > 0$. Here c_1 and c_1' are positive constants. Now, select a zero-dimensional analytic subset B_2 of $\mathbf{C}^m$ with $B_2 \cap A = \varnothing$ and such that $n_{B_2} \leqslant n_A$. Hence $\mu_{B_2} \leqslant \mu_A$. Moreover, B_2 can be taken such that $B_2 \cap E = \varnothing$ if E is any branch of $F_1^{-1}(0)$ with $C \not\subseteq A$. Then a holomorphic function F_2 exists on $\mathbf{C}^m$ such that $F_2 | A = 0$ and $F_2 | B > 0$ and such that

$$\log M_{F_2}(r) \leqslant c_2\mu_A(r+\epsilon, \epsilon) + 2\log(1+r^2) + c_2'.$$

Because $F_2(\mathfrak{z}) \neq 0$ for at least one point $\mathfrak{z}$ on each branch C of $F_1^{-1}(0)$ with $C \not\subseteq A$, this implies

$$\dim_{\mathfrak{z}} F_1^{-1}(0) \cap F_2^{-1}(0) \leqslant m-2$$

whenever $\mathfrak{z} \in F_1^{-1}(0) \cap F_2^{-1}(0) - A$. By induction, holomorphic functions $F_1, \cdots, F_m$ can be constructed such that

$$F_1^{-1}(0) \cap \cdots \cap F_m^{-1}(0) = A \cup B_{m+1}$$

where B_{m+1} is an analytic subset of $\mathbf{C}^m$ which is either empty or has dimension zero, and where $A \cap B_{m+1} = \varnothing$. Also

$$\log M_{F_p}(r) \leqslant c_p\mu_A(r+\epsilon, \epsilon) + 2\log(1+r^2) + c_p'$$

for all $r > 0$ and $p = 1, \cdots, m$. Here c_p and c_p' are positive constants. If $B_{m+1} = \varnothing$, the case is settled, if $B_{m+1} = \varnothing$ the construction is not finished, but the induction cannot be taken a step further since B_{m+1} is given and cannot be chosen. A variation of the proof of Theorem 11.1 produces a holomorphic function F_{m+1} with $F_{m+1} | A = 0$ and $F_{m+1} | B_{m+1} > 0$ such that

$$\log M_{F_{m+1}}(r) \leqslant c_{m+1}\mu_A(r+\epsilon, \epsilon) + 2\log(1+r^2) + c_{m+1}'$$

for all $r > 0$. Here c_{m+1} and c_{m+1}' are positive constants. Also $A = F_1^{-1}(0) \cap \cdots \cap F_{m+1}^{-1}(0)$.

The case of a bounded function n_A is covered by another estimate by Skoda [44] and implies that A is algebraic (compare [52]). Then $F_1, \cdots, F_{m+1}$ can be taken as polynomials. Now, minor estimates lead to Skoda's main results.

THEOREM 11.2 (SKODA [44]). *Let A be an analytic subset of $\mathbf{C}^m$ with $A \neq \mathbf{C}^m$. Let ϵ be a number with $0 < \epsilon < 1$. Then there exist holomorphic functions $F_1, \cdots, F_{m+1}$ on $\mathbf{C}^m$ and consiants $c > 0$ and $r_0 > 1$ such that*

$$A = F_1^{-1}(0) \cap \cdots \cap F_{m+1}^{-1}(0),$$

$$\log M_{F_p}(r) = c\,(\log r)^2 n_A(r + \epsilon r),$$

for all $r > 0$ *and all* $p = 1, \cdots, m + 1$.

Skoda [44] gives several variations of this result adapted to the type of growth of n_A. Also, a more exacting estimate of the constants gives applications to the theory of normal families of analytic sets.

A number of questions remain. Some of these shall be mentioned here.

1. *Question*. Is it possible to simplify Skoda's complicated proof. For instance, assume that n_A has finite order, that $0 \notin A$ and that A has pure dimension p with $0 < p < m$. Take $0 \leqslant q \in \mathbf{Z}$ with $\int_0^\infty n_A(t)\,t^{-q-2}\,dt < \infty$. Then the integral

$$V(\mathfrak{z}) = 2m \int_A \lambda_p(\mathfrak{z}, \mathfrak{x}, q)\,v^p(\mathfrak{x})$$

exists for all $\mathfrak{z} \in \mathbf{C}^m - A$. If $q = 0$ or $q = 1$, then Skoda [44] shows that V is plurisubharmonic on $\mathbf{C}^m$ and satisfies conditions (P1)–(P4). Does this remain true for each integer $q > 1$? *Added in proof.* The answer is no. See Skoda [70].

2. *Question.* Skoda [44] obtains results for domains of holomorphy in $\mathbf{C}^m$. One may ask, if the previous results extend to a connected Stein manifold M of dimension m. On M, take a strictly pseudo-convex exhaustion $\tau : M \to \mathbf{R}_+$. Here τ is of class C^∞ with $dd^c\tau > 0$ on M. For each $r > 0$, the open set $G_r = \{z \in M \mid \tau(z) < r\}$ has compact closure. If $f : M \to \mathbf{C}$ is holomorphic, define

$$M_f(r) = \operatorname{Max}\,\{|f(z)| \mid z \in \bar{G}_r\}.$$

If A is analytic and pure p-dimensional in M, define

$$n_A(r) = \int_{A \cap G_r} (dd^c\tau)^p < \infty \quad \text{if } p > 0,$$

$$n_A(r) = \# G_r \cap A < \infty \quad \text{if } p = 0.$$

Do there exist holomorphic functions $F_1, \cdots, F_{m+1}$ on M such that $A = F_1^{-1}(0) \cap \cdots \cap F_{m+1}^{-1}(0)$ and such that $\log M_{F_p}$ can be estimated in terms of n_A and invariants of (M, τ)? *Added in proof.* The answer is no. See Skoda [70].

3. *Question.* If G is an open subset of $\mathbf{C}^m$ and if $f : G \to \mathbf{C}^q$ is an open holomorphic map, then $q \leqslant m$ and each fiber of f is pure $(m - q)$-dimensional. For each point $\mathfrak{z} \in G$ the multiplicity $\nu_f(\mathfrak{z})$ of f is defined ([54] and Tung [65]). A function $\nu : \mathbf{C}^m \to \mathbf{Z}_+$ is said to be a complete intersection of dimension p, if and only if for every $\mathfrak{a} \in \mathbf{C}^m$ an open, connected neighborhood U of $\mathfrak{a}$, an open holomorphic map $f : U \to \mathbf{C}^{m-p}$ exists such that $\nu(\mathfrak{z}) = \nu_f(\mathfrak{z})$ if $\mathfrak{z} \in U \cap f^{-1}(0)$ and $\nu(\mathfrak{z}) = 0$ if $\mathfrak{z} \in U - f^{-1}(0)$. Then ν is an analytic chain of dimension p, but not each analytic chain of dimension p is a complete intersection of dimension p.

If ν is a complete intersection of dimension p, does there exist an open holomorphic map $f : \mathbf{C}^m \to \mathbf{C}^{m-p}$ such that $\nu_f = \nu$ on $f^{-1}(0)$ and such that $f = (f_1, \cdots, f_{m-p})$ where each $\log M_{f_\lambda}$ can be estimated in terms n_ν for instance by

$$\log M_{f_\lambda}(r) \leqslant c\,(\log r)^2\, n_\nu(r + r\epsilon)$$

for $r \geqslant r_0(\epsilon)$, where c is a constant?

4. *Question.* Let M be a connected Stein manifold of dimension m. Let $\tau : M \to \mathbf{R}_+$ be a strict pseudo-convex exhaustion. Let A be an analytic subset of M of pure dimension p with $0 \leqslant p < m$. Let $f : A \to \mathbf{C}$ be a holomorphic function. Define

$$M_f(r) = \text{Max}\,\{|f(z)|\ |z \in A \cap \bar{G}_r\}.$$

It is well known that there exists a holomorphic function $F : M \to \mathbf{C}$ such that $F | A = f$. Can F be chosen such that there exists an explicit estimate of M_F in terms of M_f and n_A and perhaps other invariants of A and M? In this line, Barth [4] showed that normal families of holomorphic functions on A extend to normal families of holomorphic functions on M.

References

1. L. Ahlfors, *The theory of meromorphic curves,* Acta Soc. Sci. Fenn. Ser. A No. 4 (1941), 31 pp. MR **2**, 357.

2. P. Appell, *Sur les fonctions périodiques de deux variables,* J. Math. Pures Appl. (4) **7** (1891), 157–219.

3. V. Avanissian, *Fonctions plurisousharmoniques et fonctions doublement sousharmoniques,* Ann. Sci. Ecole Norm. Sup. (3) **78** (1961), 101–161. MR **24** #A2052.

4. T. Barth, *Extension of normal families of holomorphic functions,* Proc. Amer. Math. Soc. **16** (1965), 1236–1238. MR **33** #1487.

5. S. Bergman, *Über den Wertevorrat einer Funktion von zwei komplexen Veränderlichen,* Math. Z. **36** (1932), 171–183.

6. ———, *Über eine in gewissen Bereichen gültige Integraldarstellung der Funktionen zweier komplexer Variabler,* Math. Z. **39** (1934), 76–94, 605–608.

7. ———, *Über meromorphe Funktionen von zwei komplexen Veränderlichen,* Compositio Math. **6** (1939), 305–335.

8. ———, *On zero and pole surfaces of functions of two complex variables,* Trans. Amer. Math. Soc. **77** (1954), 413–454. MR **16**, 462.

9. ———, *Bounds for analytic functions in domains with a distinguished boundary surface,* Math. Z. **63** (1955), 173–194. MR **17**, 785.

10. ———, *The number of intersection points of two analytic surfaces in the space of two complex variables,* Math. Z. **72** (1959/60), 294–306. MR **22** #2717.

11. ———, *On value distribution of meromorphic functions of two complex variables,* Studies in Math. Anal. and Related Topics, Stanford Univ. Press, Stanford, Calif., 1962, pp. 25–37. MR **27** #3829.

12. ———, *The distinguished boundary sets and value distribution of functions of two complex variables,* Ann. Acad. Sci. Fenn. Ser. AI **336/12** (1963), 28pp. MR **29** #1354.

13. J. Carlson and Ph. Griffiths, *A defect relation for equidimensional holomorphic mappings between algebraic varieties,* Ann. of Math. (2) **95** (1972), 557–584. MR **47** #497.

14. M. Cornalba and B. Shiffman, *A counter example to the "transcendental Bezout Problem",* Ann. of Math. (2) **96** (1972), 402–406. MR **47** #499.

15. H. Grauert and R. Remmert, *Plurisubharmonische Funktionen in komplexen Räumen,* Math. Z. **65** (1956), 175–194. MR **18**, 475.

16. M. L. Green, *Holomorphic maps into complex projective space omitting hyperplanes,* Trans. Amer. Math. Soc. **169** (1972), 89–103. MR **46** #7547.

17. Ph. Griffiths and J. King, *Nevanlinna theory and holomorphic mappings between algebraic varieties,* Acta Math. **130** (1973), 145–220.

18. L. Hörmander, *An introduction to complex analysis in several variables,* Van Nostrand, New York, 1966. MR **34** #2933.

19. J. King, *The currents defined by analytic varieties,* Acta Math. **127** (1971), 185–220.

20. H. Kneser, *Ordnung und Nullstellen bei ganzen Funktionen zweier Veränderlicher,* S.-B. Preuss. Akad. Wiss. Phys.-Math. Kl. **31** (1936), 446–462.

21. ———, *Zur Theorie der gebrochenen Funktionen mehrerer Veränderlicher,* Jber. Deutsch. Math. Verein. **48** (1938), 1–38.

22. R. Kujala, *Functions of finite* λ*-type in several complex variables,* Bull. Amer. Math. Soc. **75** (1969), 104–107. MR **38** #1284.

23. ———, *Functions of finite* λ*-type in several complex variables,* Trans. Amer. Math. Soc. **161** (1971), 327–358. MR **43** #7657.

24. ———, *On algebraic divisors in* C^k, Sympos. on Several Complex Variables (Park City, Utah, 1970), Lecture Notes in Math., vol. 184, Springer-Verlag, New York, 1971, pp. 223–230.

25. P. Lelong, *Sur l'extension aux fonctions entières de* n *variables, d'ordre fini, d'un développement canonique de Weierstrass,* C. R. Acad. Sci. Paris **237** (1953), 865–867. MR **15**, 416.

26. ———, *Sur l'étude des noyaux primaires et sur un théorème de divisibilité des fonctions entières de* n *variables,* C. R. Acad. Sci. Paris **237** (1953), 1379–1381. MR **16**, 123.

27. ———, *Intégration sur un ensemble analytique complexe,* Bull. Soc. Math. France **85** (1957), 239–262. MR **20** #2465.

28. ———, *Fonctions entières* (n *variables*) *et fonctions plurisousharmoniques d'ordre fini dans* C^n, J. Analyse Math. **12** (1964), 365–407. MR **29** #3668.

29. ———, *Fonctions plurisousharmoniques et formes différentielles positives,* Gordon and Breach, New York, 1968. MR **39** #4436.

30. ———, *Fonctionelles analytiques et fonctions entières* (n *variables*), Montreal Univ. Press, Montreal, 1968.

31. G. Mueller, *Functions of finite order on the ball,* Thesis, University of Notre Dame, Notre Dame, Ind., 1971, 124 pp.

32. R. Nevanlinna, *Eindeutige analytische Funktionen,* 2te Aufl., Die Grundlehren der math. Wissenschaften, Band 46, Springer-Verlag, Berlin, 1953. MR **15**, 208.

33. Y. Pan, *Analytic sets of finite order,* Math. Z. **116** (1970), 271–298. MR **43** #7661.

34. H. Poincaré, *Sur les propriétés du potential algébriques,* Acta Math. **22** (1898), 89–178.

35. L. I. Ronkin, *An analog of the canonical product for entire functions of several complex variables,* Trudy Moskov. Mat. Obšč. **18** (1968), 105–146 = Trans. Moscow Math. Soc. **18** (1968), 117–160. MR **39** #1677.

36. L. A. Rubel and B. A. Taylor, *A Fourier series method for meromorphic and entire functions,* Bull. Soc. Math. France **96** (1968), 53–96. MR **42** #509.

37. W. Rudin, *Zero-sets in polydiscs,* Bull. Amer. Math. Soc. **73** (1967), 580–583. MR **35** #1819.

38. ———, *Function theory in polydiscs,* Benjamin, New York, 1969. MR **41** #501.

39. H. Rutishauser, *Über Folgen und Scharen von analytischen und meromorphen Funktionen mehrerer Variabeln, sowie von analytischen Abbildungen,* Acta Math. **83** (1950), 249–325. MR **12**, 90.

40. B. Shiffman, *Applications of geometric measure theory to value distribution theory of meromorphic maps,* 1973, 48 pp.

41. C. L. Siegel, *Analytic functions of several complex variables,* Notes by P. T. Bateman, Institute for Advanced Study, Princeton, N. J., 1950. MR **11**, 651.

42. Y.-T. Siu, *Analyticity of sets associated to the Lelong numbers and the extension of meromorphic maps,* 1973, 9pp.

43. H. Skoda, *Solution à croissance du second problème de Cousin dans* C^n, Ann. Inst. Fourier (Grenoble) **21** (1971), fasc. 1, 11–23. MR **45** #588.

44. ———, *Sous-ensembles analytiques d'ordre fini ou infini dans* C^n, Bull. Soc. Math. France **100** (1972), 353–408.

45. W. Stoll, *Über die neuere Theorie der ganzen und meromorphen Funktionen bei mehreren komplexen Veränderlichen,* Zulassungasarbeit zur wissenschaftlichen Prüfung, Tübingen, 1949 (unpublished).

46. ———, *Mehrfache Integrale auf komplexen Mannigfaltigkeiten,* Math. Z. **57** (1952), 116–154. MR **14**, 550.

47. ———, *Ganze Funktionen endlicher Ordnung mit gegebenen Nullstellenflächen,* Math. Z. **57** (1953), 211–237. MR **14**, 970.

48. ———, *Konstruktion Jacobischer und mehrfachperiodischer Funktionen zu gegebenen Nullstellenflächen,* Math. Ann. **126** (1953), 31–43. MR **15**, 948.

49. ———, *Die beiden Haupstsätze der Wertverteilungstheorie bei Funktionen mehrerer komplexer Veränderlichen.* I, II, Acta Math. **90** (1953), 1–115; ibid. **92** (1954), 55–169. MR **17**, 893, 894.

50. ———, *Einige Bemerkingen zur Fortsetzbarkeit analytischer Mengen,* Math. Z. **60** (1954), 287–304. MR **16**, 463.

51. W. Stoll, *The growth of the area of a transcendental analytic set of dimension one,* Math. Z. **81** (1963), 76–78. MR **29** #2430.

52. ———, *The growth of the area of a transcendental analytic set,* Math. Ann. **156** (1964), 47–78, 144–170. MR **29** #3670.

53. ———, *Normal families of non-negative divisors,* Math. Z. **84** (1964), 154–218. MR **29** #2431.

54. ———, *The multiplicity of a holomorphic map,* Invent. Math. **2** (1966), 15–58. MR **35** #1832.

55. ———, *A general first main theorem of value distribution.* I, II, Acta Math. **118** (1967), 111–191. MR **36** #430a, b.

56. ———, *About entire and meromorphic functions of exponential type,* Proc. Sympos. Pure Math., vol. 11, Amer. Math. Soc., Providence, R.I., 1968, pp. 392–430. MR **38** #4706.

57. ———, *About the value distribution of holomorphic maps into the projective space,* Acta Math. **123** (1969), 83–114. MR **41** #3815.

58. ———, *Value distribution of holomorphic maps into compact complex manifolds,* Lecture Notes in Math., vol. 135, Springer-Verlag, New York, 1970. MR **42** #2040.

59. ———, *A Bezout estimate for complete intersections,* Ann. of Math. (2) **96** (1972), 361–401.

60. ———, *Deficit and Bezout estimates,* Value-Distribution Theory Part B (Edited by Robert O. Kujola and Albert L. Vitter III) Pure and Appl. Math., Vol 25, Marcel Dekker, New York, 1973.

61. B. A. Taylor, *The fields of quotients of some rings of entire functions,* Proc. Sympos. Pure Math., vol. 11, Amer. Math. Soc., Providence, R.I., 1968, pp. 468–474. MR **39** #1678.

62. P. Thie, *The Lelong number of a point of a complex analytic set,* Math. Ann. **172** (1967), 269–312. MR **35** #5661.

63. ———, *The Lelong number of a complete intersection,* Proc. Amer. Math. Soc. **24** (1970), 319–323. MR **40** #5907.

64. M. Tsuji, *Canonical product for a meromorphic function in a unit circle,* J. Math. Soc. Japan **8** (1956), 7–21. MR **21** #2051.

65. Ch. Tung, *The first main theorem on complex spaces,* Thesis, University of Notre Dame, Notre Dame, Ind., 1973, 315 pp.

66. H. Weyl and J. Weyl, *Meromorphic functions and analytic curves,* Ann. of Math. Studies, no. 12, Princeton Univ. Press, Princeton, N.J., 1943. MR **5**, 94.

Added in Proof

67. James A. Carlson, *A remark on the transcendental Bezout problem,* 1974, 11 pp. of ms.

68. James A. Carlson and Phillip A. Griffiths, *The order functions for entire holomorphic mappings,* 1974, 39 pp. of ms.

69. R. Kujala, *Generalized Blaschke conditions on the unit ball in* $\mathbf{C}^p$, Value Distribution Theory, Part A, Pure and Appl. Math., Vol. 25, Marcel Dekker, New York, 1974, 250–261.

70. H. Skoda, *Nouvelle methode pour l'etude de potentiels associes aux ensembles analytiques,* 1974, 29 pp. of ms.

Index

8480-43 P PAM
5-20

MAY 9